Agrometeorology

A Simplified Textbook

NIPA® GENX ELECTRONIC RESOURCES & SOLUTIONS P. LTD.

New Delhi-110 034

Agrometeorology

A Simplified Textbook

Prof. G. Kathiresan, Ph.D
Professor of Agronomy and Unit Head
Department of Agricultural Sciences
Agricultural Engineering College and Research
Tamil Nadu Agricultural University
Coimbatore - 641 003
Tamil Nadu

Former Dean
Anbil Dharmalingam Agricultural College and Research Institute
Trichirappalli - 620 009

Former Director
Directorate of Planning and Monitoring
Tamil Nadu Agricultural University
Coimbatore - 641 003
Tamil Nadu

NIPA® GENX ELECTRONIC RESOURCES & SOLUTIONS P. LTD.
New Delhi-110 034

NIPA® GENX ELECTRONIC RESOURCES & SOLUTIONS P. LTD.

101,103, Vikas Surya Plaza, CU Block
L.S.C. Market, Pitam Pura, New Delhi-110 034
Ph : +91 11 27341616, 27341717, 27341718
E-mail: newindiapublishingagency@gmail.com
www: www.nipabooks.com

For customer assistance, please contact
Phone: + 91-11-27 34 17 17
Fax: + 91-11- 27 34 16 16
E-Mail: feedbacks@nipabooks.com

ISBN: 978-81-96089-39-9

Composed and Designed by NIPA®.

Contents

CHAPTER-1

Agrometeorology

Food, clothing and housing are the three basic needs of a man. After that the additional requirement of a man are personal health and happiness. These can be considered as human factors. The material factors are agriculture, industry, trade, water resources management, insurance.

The continuous increase in population made the planners, researchers and cultivators to increase the food production to feed all the people. So, the second green revolution has to follow. Due to increase in air temperature and climate change, the cultivation practices to change based on the changing climatic condition. The study of climate is not only necessary for food production, but also needed for transportation, storage, value addition and consumption.

The need of the cloth for man is not only for style. The basic need is to protecting the human body from the climatic factors like heat and cold. Clothing must provide to maintain the body heat at about 37°C. Clothing must be made based on the local climatic condition. So, the traditional fiber for making the cloth is better for health.

Housing provides the internal controlled climate to the human being. It provides comfort to the inhabitants. Like clothing traditional styles, sites and methods of housing have to be followed based on the climatic condition. Though people are living in the multi-storied building, they have to concentrate on comfort in the case of sufficient aeration and ventilation. In India, the Vastu Sastrum is nothing, but it provides the needs of the human livings for their comfort with sufficient light and air.

The climatic factors for a particular day are not similar to previous day. Similarly, definitely the climate of one season is not similar to other season. Man needs some recreation in the comfort climatic areas. So, he/she needs cool climate for stay that necessitate to air conditioner particularly during summer.

AGROMETEOROLOGY

Agrometeorology is a term which is abbreviated from Agricultural Meteorology. It is also referred as Agroclimatology and has been defined in several ways. The name itself implies that it is the study of those aspects of meteorology which have direct relevance to agriculture.

The word meteorology is coined from meteor and logy. Meteor means shining body in the sky. Logy means study. Meteorology stands for science of weather. Weather is a subject which is essential for all living things in the globe. By the presence of six senses, man has a power to study the weather, whether weather is favourable for lives or changes are taken in the weather condition is leading to consequent action on the living things. How the living things can adopt for changes in the weather? Likewise he has to study by observing the things in the weather.

Development of Agricultural Meteorology

The development of meteorology in the world was taken place by the invention of certain instruments used to observe the changes taken place in the weather.

Invention	By	Year
Air thermometer	Santorio (later claimed by Galileo)	1600 AD
Rain gauge	Castelli	1639
Barometer	Evangelista Torricelli	1644 AD
Pressure tube Anemometer	Robert Hooke	1644

Robert Boyle discovered the relationship between pressure and volume in 1659.

Hadely established a relationship between trade wind and the rotation of the earth in 1735.

Benjamin Frankline studied the atmospheric electricity in 1752.

Lavosier in 1783 and Dallon in 1800 laid physical basis for meteorology.

Samuel Morse invented the electric telegraph in 1843 and this was revolutionized the weather forecasting, particularly storm warning.

The first International Meteorological Conference was held in Brussels in August 1853. It has standardized the instruments, Marine Meteorological Observation and exchange cooperation.

Societas Meteorologica Palatina (SMP) was started in 1780 at Mannheim with 39 weather observatories (14 in Germany, 4 in USA and rest in other countries). The first international compilation was made by Lamarck and the first weather map based on the data from SMP was made by H.W.Brandes in 1820 in Leipzig. W.C.Redfield prepared a series of charts of hurricanes in New York in the same period.

The first International Meteorological Congress was held in Vienna from 2-16 September 1873. Thirty two representative from 20 countries have been attended the meeting. It was a milestone in the history of international cooperation in meteorology. This congress created a permanent seven member committee. It standardized the instruction and procedures for weather observations on land and universal telegraphic weather code.

The second International Meteorological Congress was held in Rome on 14th April 1879. It was attended 40 meteorologist from 18 countries. It formed an International Meteorological Committee (IMC) with nine members by replacing seven member permanent committee. The members were non-governmental weather experts.

First conference of Directors Meteorological Services (DMS) was held in Munich on 26th August 1891. IMC worked on voluntary basis till the break of first world war 1914. It prepared and published International Cloud Atlas in 1896. Before the first world war, meteorology was in infant stage, during 1914-19 it was in adolescence stage. However, great developments were made in radio and aviation. Application of meteorology to air navigation came into act. The status of IMO and its role in international affairs was recognized until the outbreak of Second world war. Again IMO activities were subdued for five years. Conference of Directors stressed the need to convert the organization into inter-governmental. In 1935, at Warsaw Conference of Directors, commission for the application of meteorology to air navigation was replaced by International Commission for Aeronautical Meteorology (ICAeM) an inter-governmental organization. This was done due to rapid development of Civil Aviation.

In 1951, the International Meteorological Cooperation which ceased in the Vienna Congress held in 1873 was exist in 1951 as International Meteorological Organization as a non- governmental organization. It was replaced by World Meteorological Organization (WMO) which is an inter-governmental organization. It is a specialized agency of the United Nations. To give practical importance to the meteorology, the IMO was transformed to WMO.

The aims of WMO were (1) to facilitate world-wide cooperation in establishing network of weather observatories, (2) to promote the development of centers capable of providing the services of meteorology, (3) to promote rapid exchange of weather information, (4) standardization of meteorological observations and their publication, (5) to help forward the application of weather science to human activities and (6) to encourage research and training in meteorology. The conference in October 1946, signed the WMO convention by 31 countries, but it was came to force on 23rd March 1950. This date is being announced as World Meteorological Day. It had 30 members. On 19th March 1951, the first Congress of WMO convened in Paris. The membership rose to 83 countries in 1955 at the second congress held at Geneva. It further rose to 111 members at the fourth congress held in Geneva by realizing its importance.

A World Climatic Atlas was prepared for radio-tele typewriter and facsimile method used for communication purpose. In the fourth Congress of WMO held in 1963, more than 3000 voluntary weather observing ships were providing meteorological information from the world's oceans and seas. In this congress various technical notes on meteorological topics which are help to not only meteorologist but to agriculturist (Agrometeorology), aviators (Aeronautical meteorology). It provides technical assistance service with education and training to many developing countries. The advent of weather satellites (global observation), electronic computers (data exchange and processing), World Weather Watch (WWW) was recognized by United Nations General Assembly. At the beginning of the centenary year the membership of the organization rose to 136.

The WMO decided to set up three basic commissions as Commission for basic system (CBS), Commission for Instruments and Methods of observation (CIMO) and Commission for atmospheric sciences (CAS). At later five application commissions were included viz,. Commission for Aeronautical Meteorology (CAeM), Commission for Agricultural Meteorology (CAgM), Commission for Marine Meteorology (CMM), Commission for Special Application of Meteorology and Climatology (CoSAMC), and Commission for

Hydrology Meteorology (CHy). WMO had produced technical notes on guide to meteorological instruments and observing practices and published in English, French, Russian and Spanish. WMO provided the essential input of meteorological information to increase world food production, to improve shipping and marine safty, to benefit aviation and increase air safty, to improve communication networks for dissemination of meteorological information.

WMO collaborated with UN Regional Economic Commission on water resources development and management. Fifth WMO Congress which held at Geneva in 1963 approved the World Weather Watch (WWW) plan and Global Atmospheric Research Programme (GARP) in celebration with International Council of Scientific Unions (ICSU). The objective of WWW was (a) A Global Observing System (GOS) which includes observational networks and other observational facilities (weather satellite data etc.), (b) A Global Data Processing System (GDPS), which contains meteorological centres (for processing of data and archival), (c) Global Telecommunication System (GTS), which comprises of telecommunication facilities and arrangements for rapid exchange of observations and processed data, (d) A research programme (GARP) and (e) a programme in education and training.

By 1972, there were 8500 surface observations, 5500 merchant ships, plus ocean weather ships, commercial aircraft and weather satellite for global coverage of data. GDPS works through World Meteorological centres (WMC), Regional Meteorological Centres and National Meteorological Centres. In 1969 and 1970 much of meteorological aspects of agriculture was concerned to the problem of food production, that is in support of world campaign against hunger. In 1971, WMO brought out Guide to agricultural meteorological practices.

Factors Responsible Plant's Growth

Growth, development and productivity of plants depend on several factors. These factors can be broadly divided into two major groups viz., internal factors (Genetic or hereditary) and external or environmental (surrounding) factors. The environmental factors are

i) Climate (meteorological elements)

ii) Edaphic (soil)

iii) Biotic (living organisms)

iv) Physiographic (elevation)

v) Anthrophic (man)

This course on agrometeorology deals with the behaviour of the weather elements and their effect on crop production.

DEFINITIONS

Agrometeorology is a science investigating the meteorologic, climatologic and hydrologic conditions which are significant for agriculture owing to their interaction with the objects and processes of agricultural production.

A Science dealing with climatological condition which directly related or relevance to agriculture.

Meteorology

It is a branch of physics dealing with atmosphere. Atmosphere is a deep blanket of gases surrounding the earth. Meteorology is often quoted as the "Physics of the lower atmosphere". It is to study of the characteristics and behaviour of the atmosphere.

Climatology

The study of weather patterns over time and space. It concerns with the integration of day to day weather over a period of time.

Weather

The state of the atmosphere with respect to wind, temperature, cloudiness, relative humidity, pressure *etc.* at a given time.

Weather is a condition of atmosphere at a given place and at a given time. The daily variation of lower layer of the atmosphere of a smaller area like village, city or even a block or district during part of a day or complete day.

Climate

Climate is a summation of weather conditions over a given region during comparatively longer period. It is related to larger areas like zone, state, country; longer duration of time like month, season or year. It includes detailed variations extremes, frequencies, sequences of weather elements which occur from year to year, particularly in temperature and precipitation.

Agroclimatic Regions

The grouping of different physical area within the country grouped into broadly homogeneous zones based on climatic and edaphic factors.

SCOPE OF AGROMETEOROLOGY / THE NEED OF STUDY OF AGROMETEOROLOGY

For optimum crop growth, specific climatic conditions are required. Agrometeorology, thus becomes relevant to crop production; because, it is concerned with the interactions between meteorological and hydrological factors on one hand and agriculture, in the widest sense including horticulture, animal husbantary, forestry on the other.

Weather and climate are the important factors determining the success or failure of agriculture. Weather influences agricultural operations from sowing of seeds of a crop to the harvest of crops for grain and straw. Particularly rainfed agriculture depends on the mercy of the weather. In India every year there is a considerable damage by floods in one part of the country and a severe drought causing famines in another part. The total annual pre harvest losses for the various crops are estimated from 10 to 100 per cent; while, the post harvest losses are estimated between 5 to 15 per cent.

NEED OF AGROMETEOROLOGY

- The crops are to be sown at the optimum period for maximum yield. In dry lands, the time of receipt of rainfall decides the sowing date. By predicting the onset of monsoon, pre-monsoon sowing can be done.
- Study of agro-meteorology helps to minimise the crop losses due to excess rainfall, cold/heat waves, cyclones etc.
- It helps in forecasting the pest and diseases, choice of crops, irrigation and other cultural operations through short, medium and long range forecasts.
- It helps to identify places with same climatic conditions (Agroclimatic zones). This will enable to adopt suitable crop production practices based on the local climatic conditions. It also helps in the introduction of new crops and varieties which are more productive than the native crops, varieties.
- It helps in the development of crop weather models which enables to predict crop productivity under various climatic conditions.
- It helps in the preparation of crop weather calendars for different locations.

- It enables to issue crop weather bulletins to farmers.
- It enables to forecast the crop yield, based on weather, to plan and manage food production changes in a region.
- To make the farmers more "Weather conscious" in planning their agricultural operations.

DEVELOPMENT OF AGRICULTURAL METEOROLOGY

The word Climatology is derived from two Greek words, Klima + logos; Klima means slope of the earth, and logos means study. In brief, climatology may be defined as the scientific study of climate. Climatology is at once an old and a new science. Premature man was greatly affected by the phenomena of weather and climate and was unable to explain logically. Superstition served to interpret atmospheric mysteries such as rain, wind and lightning. In the early civilization, Gods were often assigned to the climatic elements, Indians still hold Ceremonial worships/ dances to Gods to produce rains at the time of drought.

The Greek philosophers showed a great interest in meteorological science. In fact the word "Meteorology" is of Greek origin, meaning, discourse or study on things above and included meteors and optical phenomena. Infact, the word "meteorology has been borrowed from Aristotle's Meteorologica" dated about 350 BC. Meteorology lasted until the beginning of the 17th century when the invention of instruments for scientific analysis of weather phenomenon was made. In 1593, Galileo constructed a thermometer and in 1643, his student Toricelli discovered the principles of mercurial Barometer. The climatological map was published by British astronomer 'Edmund Hally' in 1686. By 1800, dependable weather observations were made in Europe and USA.

An International Meteorological Organizations was established in 1878. The world meteorological organization (WMO) took its present form in 1951. It serves as a specialized agency to carryout the world wide exchange of meteorological information with the head quarters in Geneva, Switzerland.

The Indian Meteorological Department (IMD) was established in the year 1875. The division of Agricultural Meteorology was started by the IMD in 1932 to cater the needs of agriculturist and researchers. The IMD has brought out many useful publications on rainfall. The rainfall Atlas of India was published based on the rainfall data from 1901 - 1950. In addition to render advice from time to time, the IMD began to offer a regular weather service and farmers weather Bulletins in all the All India Radio stations on expected weather conditions during the next 36 hours. Weather report is also broadcast through television.

AIM, SCOPE AND IMPORTANCE

World Meteorological Organization (WMO, 1981) clearly spelled out the scope of agricultural meteorology as "The interactions between the meteorological and hydrological factors for agriculture in its widest sense including horticultural, forestry and animal husbandry". Its field of interest extends from soil layers of deepest plant and tree roots, through air layers near the ground in which crops/ trees grow and animals live, to the higher levels of atmosphere where there is efficient transport of seeds, spores, pollen and insects of economic importance. Sreenivasan (1986) stressed the role of agrometeorologist to devise a correct dynamic agronomic strategy to optimize production for sustainability. Austin Bourke (1968) emphasized the task of agrometeorologist is to apply every relevant meteorological skill to helping the farmer to make most efficient use of his physical environment with the prime aim of improving the agricultural production both quantity and quality.

It is essential to quantify the various components of biophysical environment which acts as the source of :

- radiant energy which intercepted and absorbed by the organisms in the form of dry matter accumulation through the photosynthesis process for maximization,
- soil water potential energy which is required for absorption of nutrients for the development of living cells through the optimum transpiration process,
- thermal time energy which is required for determining the rates at which plants grow and develop, and stimulates the onset of reproductive cycles in living organisms over space and time, and
- soil and aerial environment influencing the distribution and viability for insects, pathogens and parasites which survive and attack the living organisms.

CHAPTER-2

Atmosphere

Geological evidence suggests that the age of the earth is about 4,50,00,00,000 (4.5 billion) years and life began on earth between 0.6 to 1.0 billion years after earth has formed. That is life began on earth some where 4 to 3.5 billion years ago.

The plant kingdom over the globe produce oxygen about 30,00,00,000 (3 million) kg per second which is consumed by the living beings on the earth. Air which is present in the atmosphere is a mechanical mixture of gases. The various gases, solid and liquid particles in the air that envelope the earth are bound to it by the gravitational attraction. This complex content of gases, solid and liquid particles is called atmosphere. The atmosphere extends above the surface of the earth to great heights. There is no sharp boundary between the atmosphere and the outer space. For our convenience, the height of the atmosphere is taken as 1000 km. The lowest one km which is called planetary boundary layer, contains 10 per cent of the mass of the atmosphere. All biological and human activities are confined to this planetary boundary level. The mass of the atmosphere is about 5.6 x 1018 kg (mass of earth is 6 x 1024 kg). The mass of the oceans water is about 1.4 x 1021 kg. This shows that the mass of the oceans is more than 250 times the mass of the atmosphere. The density of the dry air at the earth surface is 1.225 kg/m^3 and at an altitude of 5.5 km this drops to 0.66 kg/m^3 and at 30 km altitude it is about 0.013 kg/m^3.

STRATIFICATION AND COMPOSITIONS OF ATMOSPHERE

The atmosphere is a mechanical mixture of many gases, not a chemical compound. In addition, it contains water vapour, (4 per cent atmospheric composition) and huge number of solid particles, called aerosols. Some of the gases (N, O, Ar, CO_2) may be regarded as permanent atmospheric components that remain in fixed proportions to the total gas volume. Other constituents vary in quantity from place to place and from time to time. If the suspended particles, water vapour and other variable gases were excluded from the atmosphere, we would find that the dry air is very stable all over the Earth up to an altitude of about 80 kilometers.

Composition of Atmosphere

Table : Principal gases comprising dry air in the lower atmosphere.

Constituent	Percent by volume
Nitrogen (N_2)	78.08
Oxygen (O_2)	20.94
Argon (Ar)	0.93
Carbon dioxide (CO_2)	0.03
Neon (Ne)	0.0018
Helium (He)	0.0005
Ozone (O_3)	0.00006
Hydrogen (H_2)	0.00005
Krypton (Kr)*	Trace
Xenon (Xe)*	Trace
Methane (Me)	Trace

• Chemically inert, never found in any chemical compounds.

As shown in the table, gases, nitrogen and oxygen, make up about 99 percent of the clean, dry air. The remaining gases are inert and constitute about 1 per cent of the atmosphere, generally homogeneous and it is called as homosphere. At higher altitudes, the chemical constituents of air get changed considerably. This layer is known as the heterosphere.

i) Gases

Nitrogen (N_2)

Relatively inactive, main function is to regulate combustion by diluting O_2. Indirectly helps oxidation (mainly diluents).

Carbon dioxide (CO_2)

Plants take CO_2 in the process of photosynthesis, and it efficiently absorbs the heat from upper atmosphere as well as the Earth. This emits half of the absorbed heat back to the Earth. Influences flow of energy through the atmosphere. The proportion remains the same; but, percentage increases due to burning of fossil fuels. From 1890 to 1970 - CO_2 content has been increased more than 10 times - warming of lower atmosphere - climatic changes.

Ozone (O_3)

It is a type of oxygen molecule formed by three atoms rather than two. It is found only in very small quantity in the upper atmosphere. It is less than 0.00006 per cent by volume. The maximum concentration of ozone is found between 30 and 60 km. Although it is formed at higher levels and transported downward. It is the most efficient absorber of the burning ultraviolet radiation from the Sun, acting as a filter. Absence of ozone layer will make the Earth's surface unfit for human habitation - for all living organisms. Of all the gases, oxygen happens to be the most important for all living organisms.

i) Ozone layer

The earth's atmosphere is composed of numerous gases. The main gases are nitrogen and oxygen, which makes up 78 per cent and 21 per cent of the volumes of air respectively. The remaining one per cent of the atmosphere gases is made up of trace-gases, which include the green house gases - carbon di oxide, water vapour and ozone, so-called because they are involved in the earth's natural green house effect which keeps the planet warmer than it would be without an atmosphere. The trace gas ozone is important in protecting life on earth. Ozone is a colourless, odourless gas. It performs an important function of filtering out the harmful UV rays from the sun's radiation.

The International day for the preservation of the ozone layer has been proclaimed on 16th of September by U.N. General Assembly.

What is the Need to Protect the Ozone Layer?

The atmosphere makes life possible on the Earth/ Planet, because it has three very important functions. They are (1) It contains life-giving oxygen, (2) it keeps the earth warm and (3) it absorbs the deadly ultraviolet (UV) rays of the sun's radiation coming from atmosphere.

Ozone is continuously produced and destroyed naturally. However, chlorofluorocarbons (CFCs) released by human activities reach the stratosphere and destroy ozone. These are called Ozone Depleting Substances (ODS). CFCs and are used in refrigeration, air conditioning, fumigation, making foams, aerosols, as fire fighting agents, cleaning agents, solvents and so on.

British scientists while measuring the atmospheric ozone over the America in the late 1980s made in unpleasant discovery as ' there was much depletion in the ozone layer over Antartica each spring. Such areas are referred to as "Ozone hole". In 1985, nations agreed in Vienna to take "appropriate measures to protect human health and the environment against adverse effects resulting or likely to result from human activities, which modify or are likely to modify the Ozone layer". Thus the Conventions for the Protection of the Ozone layer was born. In September 1987, an agreement was reached on specific measures to be taken resulting in the Montreal Protocol on Substances that Deplete the Ozone layer signed. In 1994, the United Nations General Assembly proclaimed September 16, the International Day for the preservation of the ozone Layer, commemorating the date of the signing in 1987, of the Montreal Protocol. India has ratified the Vienna Convention in 1991 and the Montreal Protocol in 1992. These multilateral agreement provide technical and financial assistance to move on to using Ozone friendly replacement.

Individual responsibility on saving of ozone layer:

As consumers, we can be more aware and check whether the products,

(1) We buy the product which likely to contain CFCs. Where we have a choice, we could choose ozone friendly products, e.g., conventional spray pumps for insecticides, rather than aerosols.

(2) While buying a refrigerator or air conditioner, choose one with CFC-free technology. While getting a refrigerator which has CFC gas in the compressor repaired, ensure the gas is not released into the atmosphere.

(3) We can as far as possible, buy and use pillows and mattresses stuffed with cotton instead of foam. Let us not use Styrofoam glasses and cups,

(4) Use leaf or reusable plates. Let's do our bit to save the ozone layer.

ii) Water vapour

Water vapour is one of the most variable gases in the atmosphere, which is present in small amounts, but it is very important. The water vapour content of air may vary from 0.02 per cent by volume in a cold dry climate to nearly 4 per cent in the humid tropics. The variation in the percentage over time and place are very important considerations climatically. Like CO_2, water vapour has insulating action of the atmosphere. It absorbs not only the wave terrestrial radiation, but also a part of the incoming solar radiation. Thus it regulates energy transfer through the atmosphere. Water vapour is the source of all clouds and precipitation.

iii) Dust particles

Dust particles are major contributory factor in the formation of clouds and fogs. It is responsible for the red, orange colour of the sky at Sun rise and sunset.

LAYERED STRUCTURE OF THE ATMOSPHERE

During the international Geophysical year (1957-62), important discoveries were made about the atmosphere and many new facts came to light. The Earth's atmosphere consists of zones or layers arranged like spherical shells according to altitude above the Earth's surface.

According to Peterson, the atmosphere is divided into the following, more significant, spheres. Figure below shows different sphere of the atmosphere according to altitude and changes of temperature.

i) Troposphere

ii) Stratosphere

iii) Ozonosphere (also called Mesosphere)

iv) Ionosphere

v) Exosphere

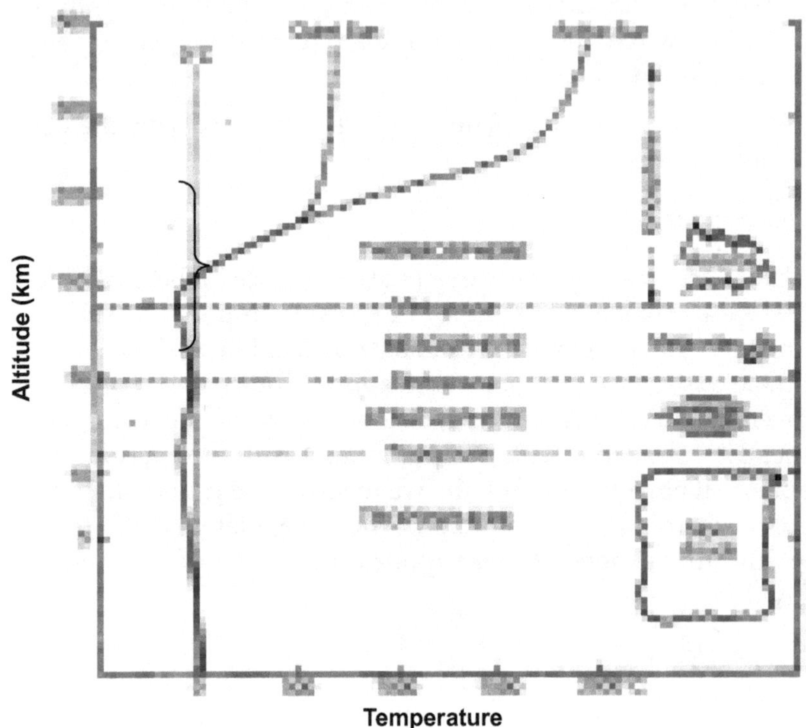

Troposphere

It contains about 75 per cent of the total gaseous mass of the atmosphere. It has been derived from the Greek word "tropos" meaning "mixing" or turbulence. Under normal conditions, the height of the troposphere at the poles is about 8 km while at the equator it is about 16 km. Troposphere is marked by turbulence and eddies. It is also called connective region. Various types of clouds, thunderstorms as well as cyclones and anticyclones occur in this sphere, because of the concentration of almost all the water vapour (4 per cent of the atmospheric composition) and aerosols in it. Wind velocities increase with height and attain maximum at the top.

Stratosphere

Stratosphere is a layer of atmosphere which lies above the tropopause. The troposphere and stratosphere are separated by a narrow layer called tropopause. It lies beyond the height of 8 to 18km extending up to 50km depending on latitude. It is dust free, cloudless and warmest layer.

The lower stratosphere is isothermal in character (16-30 km). There is a gradual temperature increase with height beyond 20 km i.e. upper stratosphere. No visible weather phenomena occur above tropopause.

Ozonosphere or Mesosphere

Mesosphere lies above the stratosphere. The stratosphere and mesosphere are separated by a narrow transitional layer up to 8-9 km called stratopause.

There is maximum concentration of ozone between 30 and 60 km above the Earth's surface. Because of the concentration of ozone in this layer it is called the ozonosphere. It is a warm layer, because selective absorption of ultra violet radiation by ozone. In fact, it acts as a filter for ultra violet radiation from the Sun.

In this layer, the temperature increases with height @ 5° C/Km. The maximum temperature recorded in the ozonosphere is higher than at the Earth's surface. Because of the preponderance of chemical processes, this sphere is sometimes called as chemosphere.

Ionosphere

Ionosphere, according to Peterson, lies beyond the ozonosphere at a height of about 60 km. above the Earth's surface. At this level the ionization of atmosphere begins to occur. Above ozonosphere, the temperature falls, again reaching a minimum of about 100° C at a height of 80 km above the Earth's surface. Beyond this level the temperature increases again due to the absorption of short - wave solar radiation.

Layers of Ionosphere

D Layer	-	60-90 km.
E Layer	-	90-130 km
Sporadic Layer	-	110 km.
F2 Layer }	-	150 km.
F1 }	-	150-380 km
G Layer }	-	400 km and above

Exosphere

The outermost layer of the Earth's atmosphere is known as the exosphere which lies between 400 and 1000 km. At such great height, density of atoms in the atmosphere is extremely low. Hydrogen and helium gases predominate in the outermost region. Kinetic temperature may reach 5568 ° Celsius.

CHAPTER-3

Weather and Climate

In climatology, terms "Weather" and "Climate" have different connotations. Weather refers to the state of atmosphere at any given time denoting the short- term variations of atmosphere in terms of temperature, pressure, wind, moisture, cloudiness, precipitation and visibility. Weather is highly variable. It is constantly changing, sometimes from hour to hour and at other times from day to day. The aforementioned properties of the atmosphere are subject to constant change and their state at any time determines the state of the weather. However, weather elements are not separate rather they are closely related with each other.

David I. Blumenstock, an eminent climatologist has defined the weather in more restrictive sense. According to him, weather is the behaviour of the lower atmosphere with particular emphasis on the atmospheric behaviour which affect the lands and oceans and have a marked influence on the organisms that live upon the lands, within the waters of the earth or in the lower air. For the geographer, this definition of weather is more satisfactory. Since, he is interested more in the working of the lower atmosphere which affects the life of man as well as natural environment directly. Thus, we come across a great many varieties of weather covering a wide range of conditions.

CLIMATE

Climate on the other hand, is the sum, total, of the variety of weather conditions on the area or a place from day to day. Thus, climate may be defined simply as average weather, the term climate denotes a description of aggregate weather conditions. True climate describes an area's average weather, but it also includes common deviations from the average as well as the extreme conditions. For getting the vital climatic information the range and frequency of extremes are of great significance. Therefore, climate may be defined as the sum of all statistical weather information's of a particular area during a specified interval of time, usually several decades. The world Meteorological Organizations has suggested standard period of 31 years for calculating the climatic averages of different weather elements. Different climatalogists have defined the climate. But all are based on weather, over a period of time.

Weather parameters / elements:

Different combinations of weather elements make up the climate of a particular place or area. The following are some of the most important weather elements.

1. Solar radiation
2. Temperature
3. Air pressure
4. Wind Velocity and Wind Direction
5. Moisture (Humidity)
6. Cloudiness (Sunshine hours)
7. Precipitation (Rainfall)

All these are highly variable and constitute the weather /climate. A change in one of the elements generally brings about changes in the others.

Factors Affecting the Weather and Climate : (Climatic Control)

1. Latitude

The distance from the equator either south or north, largely create variations in the climate. Based on the latitude the climate has been classified as (i) Tropical (ii) Subtropical (iii) Temperate and (iv) Polar

The tropical climate is characterized by high temperature throughout the year. Subtropical is also characterized by high temperature alternating with low temperature in winter. The temperate climate has low temperature throughout the year. The Polar climate is noted for its very low temperature throughout the year.

2. Altitude (Elevation)

The height from the mean sea level creates variations in climate. Even in the tropical regions, the high mountain have temperate climate. The temperature decreases by 1.8°C for every 300 m from the sea level. "Generally there is a decrease in pressure and increase in precipitation and wind velocity. The above factors alter the kind of vegetation, soil types and the crop production.

3. Precipitation

The quantity and distribution of rainfall decides the nature of vegetation and the nature of the cultivated crops. The crop region are classified on the basis of average rainfall which are as follows.

Climatic region based on rainfall

Rainfall (mm)	Name of the Climatic Region
Less than 500	Arid
500 - 750	Semi Arid
750- 1000	Sub-humid
More than 1000	Humid

Soil is a product of climatic action on rocks as modified by landscape and vegetation over a long period of time. The colour of the soil surface affects the absorption, storage and ^eradiation of heat. White colour reflects while the black absorbs more radiation. Due to differential absorption of heat energy, variations in temperature are created at different places. In black soil areas the climate is •hot while in red soil areas, it is comparatively cooler due to lesser heat absorption.

4. Nearness to Large Water Bodies

The presence of large water bodies like lakes and sea affect the climate of the surrounding areas, eg. Islands and Coastal areas. The movement of air from earth surface and from water bodies to earth modify the climate. The extreme variation in temperature during summer and winter is minimized in coastal areas and Islands.

5. Topography (Relief)

The surface of landscape (leveled or uneven surface areas) produces marked changes in the climate. This involves the altitude of the place, steepness of the slope and exposure of the slope to light and wind.

6. Vegetation

Kind of vegetation characterize the nature of climate. Thick vegetation is found in tropical regions where temperature and precipitation are high. General types of vegetations present in a region indicates the nature of climate of that region.

7. Other Factors are

i) Semipermanent high and low pressure systems

ii) Wind and air masses

iii) Atmospheric disturbances or storms

iv) Ocean currents

v) Mountain barriers.

CHAPTER - 4

Solar Radiation

The Sun is the primary source of heat to the Earth and its atmosphere. The heat received from other celestial bodies as well as the interior of the Earth is rather too significant. The distance that separates the Earth from the Sun is about 149.0 million km. The diameter of the Sun measures roughly about 1.3824 million km. The surface temperature of the Sun is estimated between 5500° C and 6100° C. (or 5762 ° K).

The weight of the sun is about 1.9 X 10 30kg. The radius is 6.97 X 10 8 m or the diameter of the sun is 14 X 10 5 km. The mass of the earth is 6 X 10 24 kg, the radius is 6.4 X 10 6 m. The gravitational force of the earth is 9.81m/s^2. Whereas the sun's attractive force is 274m/s^2. This is 25 times higher than earth's gravitational force.

In the sun, very hot materials electrons which is negatively charged particles are in continuous motion at high speed, collide with one another and cause vibrations. This gives rise to electromagnetic waves. Very high speed particles vibrate at high speed produces short waves. Slower particles collision produces long waves. In the sun, the gases which are at high temperature i.e., more than 6000°K, produce electrically charged particles with different speeds. As a result of their collision, electromagnetic waves of different wave length are emitted by the sun. These waves, constitute the electromagnetic spectrum of the sun which extends from gamma rays (λ=10-13m)to radio waves (λ=103m). The range is 1016m. However about 99 per cent of the solar radiation lies between (λ=0.15 - 4.0 μm).

Solar radiation provides more than 99.9 per cent of the energy received by the Earth. Undoubtedly, the radiant energy from the Sun is the most important control of our weather and climate. The most astonishing fact about the incoming solar radiation (insolation) that strikes the Earth's surface is that it is equal to about 23 billion horse power. Actually this amount of energy received from the Sun that acts as the driving force for all the atmospheric as well as biological processes on the Earth. Besides, all other sources of energy found on Earth such as coal, oil and wood etc. are nothing but converted form of solar energy.

The word 'insolation' is contraction of incoming solar radiation. Radiant energy from the Sun that strikes the Earth is called insolation.

All the radiation travels with speed of $3X10^8$ m/s in vacuum. According to Wien's law sun emits energy at average surface temperature of 6000° K and the earth at its average surface temperature of 288° K. That is the sun emits maximum amount of radiation energy at wave length λ max = 0.5 μm. The earth emits maximum amount of radiation energy at wave length λ max = 0.5 μm.

All objects in the universe emit radiant energy as long as their temperature is more than 0° K, but ceases when it is less than 0° K. as the temperature of the object increases it sends more and more energy into space and obeys Stefan Boltzman law which gives $R \div T^4$, where, R=maximum rate of radiation emitted per unit surface area of the object, T= temperature of the object in degrees Kelvin.

When an object emits and receives energy at the same rate, it is said to be in radioactive balance with its surroundings. If it is not in radiative balance either it receives or loss energy depending on its temperature less than or more than its surroundings. By this the object get warms up or cools down.

Radiant energy travels in space/vacuum in the form of electromagnetic waves. Radiant energy waves posses both electric and magnetic properties; consequently they are called **electromagnetic waves**. Electromagnetic waves travel in all direction away from the source with the speed of light ($C = 3 X 10^8$ m/s). The wave length ë and frequency f are related by the formula $C = f\lambda$.

In the insulation light energy, about 9 per cent lies in UV-radiation (0.15μ m-0.38μ m): This causes photochemical effects, bleaching and sunburn etc., 44 per cent lies in visible light (0.38μm-0.7μm i.e.,violet to red), 46 per cent lies in IR (0.7 to 2.3μm): This causes radiant heat with some chemical effects.

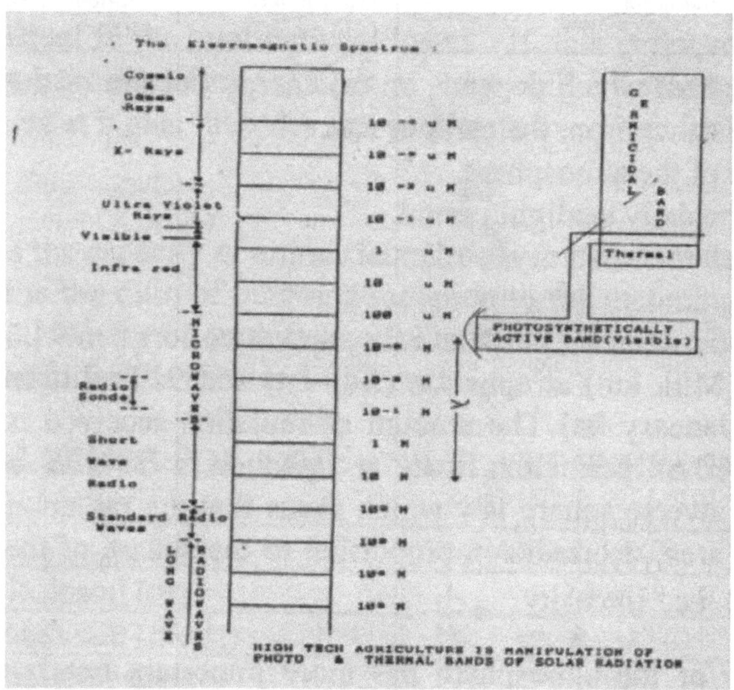

Human eyes are sensitive to waves of certain wave length, which are called light waves. Radiant energy waves which travel through space/vacuum as light waves is called **visible radiation**. Visible radiation spectrum consists of VIBGYOR light (Violet, Indigo, Blue, Green, Yellow, Orange and Red). The average wave length of these spectrum light as follows.

S.No.	Spectrum light colour	Wave length (µm)
1.	Violet	0.400
2.	Indigo	0.424
3.	Blue	0.490
4.	Green	0.575
5.	Yellow	0.585
6.	Orange	0.647
7.	Red	0.710

The electro-magenetic spectrum consists of Gamma rays, X-rays,ultraviolet (UV) radiation, Visible light, Infrared (IR) radiation, microwaves, short and long waves. The wave length (λ) and frequency (f) of these waves are as follows.

S.No.	Radiation	Wave length (λ in meters)	Frequency (Hz/s)(f in Hertzs/second)
1.	Gamma rays emanates in nuclear explosions	$<10^{-10}$	$\geq 10^{18}$
2.	X - rays (medical uses)	10^{7}	$5 X 10^{18} - 10^{19}$
3.	UV - light	$0.3 X 10^{-6}-10^{-8}$	$10^{18} - 2 X 10^{18}$
4.	Visible light / violetRed (rain bow part)	$0.4 X 10^{-6} 0.8 X 10^{-6}$	$3.7 X 10^{14} 10 X 10^{14}$
5.	Principal solar radiation (UV, visible and IR)	$10^{-7} - 10^{-6}$	$10^{15} - 10^{13}$
6.	IR (thermal radiation) Max. radiation at room temperature	$0.8 X 10^{-6} - 10^{-4}$ $8 x 10^{-6}$	$3.7 X 10^{14} - 10^{12}$ $0.3 X 10^{14}$
7.	Microwaves or very short radio waves	$10^{-4} - 1$	$10^{12} - 10^{8}$
8.	UHF (Ultra High Frequency) TV channels 14 – 83	0.1 - 1	$3 X 10^{9} - 3 X 10^{8}$
9.	VHF (TV channels 2-13) FE – Radio	1 - 10	$3 X 10^{8} - 3X 10^{7}$
10.	Short Radio Waves	10 - 100	$3 X 10^{7} - 3 X 10^{8}$
11	Long Radio Waves	200 – 600	$5 X 10^{6} - 1.5 X 10^{5}$

The fusion furnace of the sun converts about 4 million tons ($4 X 10^{9}$ kg) of hydrogen into helium and radiates an amount of energy $3.8 X 10^{23}$ Kw/s as a by product. The solar energy incident on $1m^2$ area held normal to the sun's rays at the outer boundary of the atmosphere in one second is $1.38 X 10^{3} W/m^{2}$. It is virtually constant and called **solar constant**.

A house with peak load requirement of 2.5 KW would require a solar collector of 3.6 m^2 with 100 per cent efficiency. The total solar energy received at the earth in one day is equivalent to 35 lakhs nuclear explosions or 10000 hurricanes or 10^{8} thunderstorms or 10^{11} tornadoes.

According to one estimate, the solar radiations received at the outer boundary of the atmosphere are about $1.74 X 10^{14}$ KW. Of this about (30 per cent) $5.2 X 10^{13}$ KW is reflected back as shortwave radiation. $8.2 X 10^{13}$ KW (about 47 per cent) is directly absorbed by the land, oceans and atmospheric system and the remaining $4.0 X 10^{13}$ KW (about 23 per cent) is utilized in driving the hydrologic cycle through evaporation, convection and precipitation. About $3.7 X 10^{11}$ KW (about 21 per cent) of the incident solar energy is utilized (as input into Biosphere) for driving winds, waves, convection and sea currents. $4 X 10^{9}$ KW (0.023 per cent of the incident solar energy) is used in photosynthesis process on earth. The green canopy of plant kingdom consumes solar energy about (0.01 ly/min) 7 watt/m^2. The total solar energy intercepted by the earth is 2.55 X 10^{18} Cal/min or 3.67 X 10^{21} Cal/day.

Solar energy per unit area is expressed in Langley (ly) or Kilo-langley. 1 ly = 1 Cal/cm^2.

LIGHT

Solar radiation consists of a bundle of rays of radiant energy of different wave lengths. The visible portion of the solar spectrum appears as light. Light travels with a speed of 2.976 lakhs km/sec. and it takes 8 minutes to reach the Earth.

Light is the total effect of the combination of the seven different colours, namely red, orange, yellow, green, blue, indigo and violet (VIBGYOR). The waves that produce the effect of red colour are the longest and those producing the violet are the shortest in wave length. Waves shorter than the violet are called ultra-violet rays, while those longer than the red are known as infra-red rays. Percentages of solar constant in the main spectral bands are:

The ultra-violet rays form only 6 per cent of the insolation, but have strong photochemical effects on some substances. The infra-red rays, even though invisible, form 43 per cent of the insolation. They are largely absorbed by water vapour that is concentrated in the lower atmosphere.

The principle effects of radiation are of thermal, photochemical, germicidal and photoperiodic nature. Thermal and photochemical effects are of major importance in determining the growth rate, biomass production and quality of agricultural produce. Therefore, efficient use of solar energy is an important aspect in agro-meteorology to utilize vast available natural resources. Solar radiation, available at the earth's surface, provides an index to compare the vegetation productivity under natural agroecological conditions when soil water and nutrients are not limited.

In agriculture, solar energy is conserved, through its fixation in biomass by the process of photosynthesis where carbon dioxide from air is converted into carbohydrates expressed as

$$6CO_2 + 12H_2O \xrightarrow[\text{(Photosynthetically Active Radiation-PAR)}]{\text{Sunlight}} (CH_2O)_6 + 6O_2 + 2H_2O \quad \text{(Photosynthesis)}$$

$$(CH_2O)_6 + 6O_2 \longrightarrow 6CO_2 + 6H_2O + \text{Chemical energy}$$

(Respiration)

About 40 to 50 per cent of carbohydrates (weight) produced are lose through respiration by providing the chemical energy for the plant processes. Remaining carbohydrates are distributed into different plant organs for their development. This is a dynamic process with formation of complex material and its simultaneous break down to meet the energy requirements for all the growth process.

The quantification of efficiency parameters in terms of the amount of solar radiation, growth rates parameters and solar energy conversion parameters into biomass will provide an insight to further improve these productivity parameters for reducing the gap between actual and potential/maximum possible productivity levels.

PROCESSES OF HEAT ENERGY TRANSFER

i. Radiation

Radiation is the process of transmission of energy by electromagnetic waves and means that the energy emitted by the Sun reaches the Earth. It does not require medium. Transfer of energy by radiation occurs over a wide spectrum of wavelengths ranging from very short X-rays to very long radio waves. Solar radiation consists of stream of flow of particles and they are called Qunita or Photons.

ii. Conduction

Conduction is the process of heat transfer through matter by molecular activity. In this process heat is transferred from part of a body to another or between two objects touching each other. Conduction occurs through molecular movement.

iii. Convection

Convection is the process of transfer of heat, through movement of a mass or substance from one place to another. Convection is possible only in gases or fluids, for those have internal mass motions. In solid substances this type of heat transfer is impossible. Convection is much faster than conduction.

HEAT BUDGET

Of the total solar radiation reaches the outer limit of the atmosphere, about 26 per cent is reflected by clouds or scattered back to space by suspended particles and it is not used to heat the sir. The Earth's surface reflects 4 per cent of radiation to the space. About 19 per cent of solar radiation is absorbed by gases and water vapour, about 24 per cent is absorbed by the Earth from scattering of clouds and atmosphere. Thus approximately two-thirds of the total radiation is effective in heating the Earth. Energy balance of the Earth and its at most atmosphere is given in Figure.

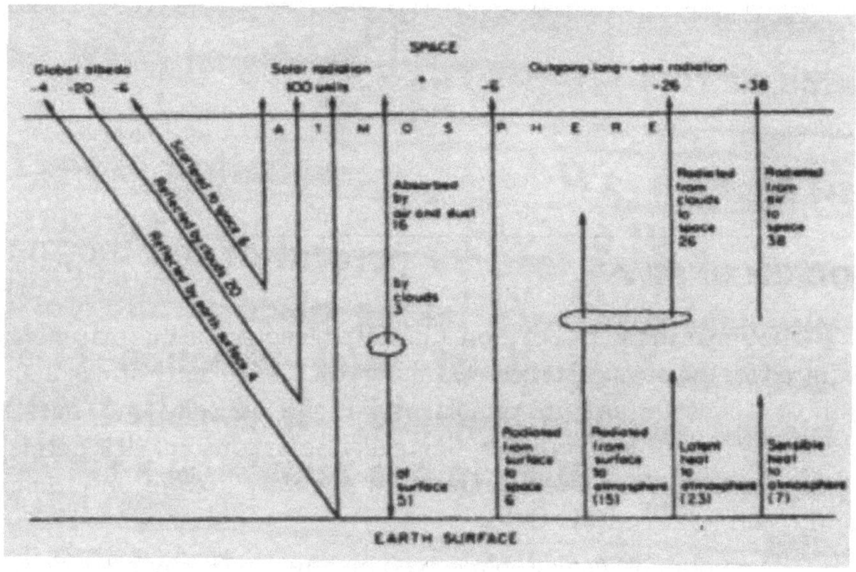

The total energy coming to the Earth over a considerable period of time is equal to the total outward losses. If, this is not so, the Earth would seen become either very hot or very cold. Actually there is a deficit of heat at higher latitudes and surplus in low latitudes.

ALBEDO

It is the capacity of any surface to reflect the incoming radiation (light) or it is the ratio of outgoing radiation to the incoming radiation. The total reflectivity of Earth is known as **Earth's albedo**. The average value of albedo for the Earth is 34 per cent.

$$\text{Albedo of a surface} = \frac{\text{Global radiation from the sun and sky reflected by the surface}}{\text{Global radiation from the sun and sky incident on the surface}}$$

Albedos for Soil and Plant Surface

Type	Albedo (%)
Black soil (dry)	14
Black soil (moist)	8
Gray soil (dry)	25 - 30
Gray soil (moist)	10 - 12
Ploughed field (moist)	14 - 17
Pearl millet	24
Moong	26
Wheat	23 - 25
Rice	12
Potato	19
Lucerne	23 - 32

SOLAR CONSTANT

The Sun is the source of more than 99 per cent of the thermal energy required for the physical processes taking place in the Earth atmosphere. Almost a constant amount of solar radiation (1.43 cal / cm^2/min) is emitted by the Sun continuously. It is called **solar constant**. It is defined as the energy falling in one minute on a surface area of one square centimeter.

The amount of insolation received on any date at any place on the Earth is favoured by,

i) The solar constant which depends on (a) energy output of the Sun and (b) distance from the Earth to Sun.

ii) Transparency of the atmosphere.

iii) Duration of the daily sunlight period.

iv) Angle at which the Sun's rays strike the Earth.

The distance between the Earth and the Sun varies between 94.5 million miles (157.5 Mill. km) at aphelion (July I^{st}) and 91.5 million miles at perihelion (January I^{st}). The amount of radiation received is seven per cent greater at perihelion than at aphelion. This is a consequence of the inverse square law which states that the radiation received on any unit area, decreases in proportion to the square of the distance to the source ie., Intensity a =$1/d^2$

Transparency of the atmosphere has more important bearing upon the amount of insolation which reaches the Earth's surface. The areas having heavy dust, clouds, water vapour and cloudiness or polluted air will receive less, direct, insolation. The transparency of atmosphere depends on the latitude of a place. At middle and high latitudes the Sun's rays must pass through thicker layers of reflection scattering materials and it is not so at tropical latitudes.

Solar Constant (S)

The amount of solar radiation (energy) incident on unit area in unit time on a surface held at right angles to the solar beam at the outer boundary of the atmosphere.

$$\text{Solar constant (S)} = \frac{56 \times 10^{26}\,\text{Cal / min}}{4\pi(1.5 \times 10^{13}\,\text{cm})^2}$$

On an average the mean cloud coverage of the earth is slightly more than 50 per cent. Under this condition the distribution of insolation (incoming solar radiation) is as follows.

Reflection and back scattering 35% (about 92 kly/yr)

Absorption by ozone 2% (5 kly/yr)

Absorption by clouds, water vapour, dust etc. 20% (53 kly/yr)

Absorption by the earth's surface 43% (113 kly/yr)

The earth atmosphere together absorbes 65% (171 kly/yr)

Solar radiation is the primary electromagnetic spectrum showing the wave length of different types of radiations is shown below.

Type of radiation	Wave length (micron)
1. Cosmic rays	10^{-7} to 10^{-4}
2. Gamma rays	10^{-7} to 10^{-3}
3. X-rays	10^{-3} to 10^{-1}
4. UV rays	1 to 390
5. Visible rays	390 - 760
6. Infra red rays	760 - 10^{6}
7. Radio wave rays	10^{6} to 10^{13}

ENERGY CONSERVATION IN ATMOSPHERE

Atmosphere exerts a force on the earth's surface due to its weight and therefore the average atmosphere pressure at sea level value of 1.013×10^5 Pa or 1013.2 mb at absolute temperature of 273^0 K. Vertical variability of pressure and density of air in atmosphere is more rapid than horizontal variability, however in the lower atmosphere of 100 km height, pressure and density of atmosphere decreases with height. Atmosphere pressure is assessed as weight of air column of unit cross sectional area of height.

Gas Laws of Atmospheric Air

Where, R is gas constant for one kg of gas and T is temperature in ^{0} K, if p_a is density of air, then gas equation takes this from as

$$P = p_a RT$$

This expression is used at different atmosphere conditions.

i. at constant volume, P is proportional to T,

ii. at constant pressure, V is proportional to T,

iii. at constant temperature (isothermal) P is inversely proportional to V,

iv. at constant energy (adiabatic) P,V and T may change.

RADIATION LAWS

Sun, the main source of energy, emits solar energy in the form of electromagnetic radiation. These raditions are regulated by the wave equation ($c=v\lambda$) where c is velocity of light ($c=3 \times 10^8$ ms^{-1}) and v frequency of wave motion and λ is wavelength of radiation emitted.

RADIATION BALANCE

Radiation balance over vegetations is important as radiation absorption and transmission are required for a number of physiological principles. The net radiation flux, R_n, is the balance between the net incoming shortwave (R_{sw}) and net outgoing longwave(R_{Lw}) radiation at the thermal equilibrium surface. Considering unit area of any vegetative/animal/base soil/water surface, the radiation balance expression is.

$$R_n = R_{sw} - R_{Lw}$$

Net radiation is the useful component varying between 50 to 80 per cent of daily radiation load.

It is assumed as sign convention that all radiations coming to the surface are taken as positive, whereas those leaving the surface are taken as negative. In explicit form it can be expressed as

$$R_n = (1-\alpha_b)\ R_{st} + \epsilon_\alpha\ R_{Lwd} - \epsilon_b\ R_{Lwd}$$

During clear sky days R_n / R_{st} ratio is almost constant as long wave radiation emitted from earth's surface also increased with increase of incoming solar radiation. However during cloudiness, the ratio R_n / R_{st} decreases as both long wave components are of the some order. During night time R_n becomes negative as only balance of long wave components exist in nature.

The instantaneous radiation balance over a short green vegetative grass surface is expressed as

$$R_n = (1-\alpha_g)\ R_{st} + R_{Lwd} - \epsilon\sigma T_1^{\ 4}$$

Where, α_g and T_1 are reflection coefficient and radiative temperature of green lawn surface.

If a green vegetative leaf is lying on the lawn surface, then radiation balance will be expressed as

$$R_n = (1-\alpha_l)\ (1+\alpha_g)\ R_{st} + \epsilon_\alpha\ R_{Lwd} + \epsilon_\alpha\ R_{Lwe} - 2\epsilon\sigma T_2^{\ 4}$$

Where, α_e and T_2 are reflection coefficient and temperature of leaf surface exposed on the lawn surface. The leaf will receive an additional shortwave radiation of $p_g\ R_{st}$ and longwave radiation from environment of vegetative lawn surface ($\epsilon R_{Lwe} = \epsilon\sigma T_1^{\ 4}$). Both sides of leaf surface will emit long wave radiation. R_n is net radiation per unit of total leaf area i.e., twice the leaf area index.

Similarly if a tree is planted in the lawn, then it will also receive additional radiations reflected and emitted by the lawn surface assuming its cylindrical shape

$$R_n = (1-\alpha_c)\ (1+\alpha_g)\ R_{st} + \epsilon_\alpha\ R_{Lwd} + \epsilon_e\ R_{Lwe} - 2\sigma T_3^{\ 4}$$

Where, α_c and T_3 are shortwave reflectivity and mean temperature of tree body surface.

Net Radiation (R)

Net radiation is the balance of incoming and outgoing radiation at a place on a unit area surface. It is a dominant term in energy balance equation and follows the similar pattern of incoming solar radiation during daytime, however during nighttime it becomes negative as longwave radiations emitted by the earth or vegetative surface, is the dominant term. Diurnal pattern of net radiation exists. Net radiation close to zero near sunrise and after sunset. If advection components is small then net radiation is close to the total available heat energy (H). The convention of recording data is as follows with different parameters.

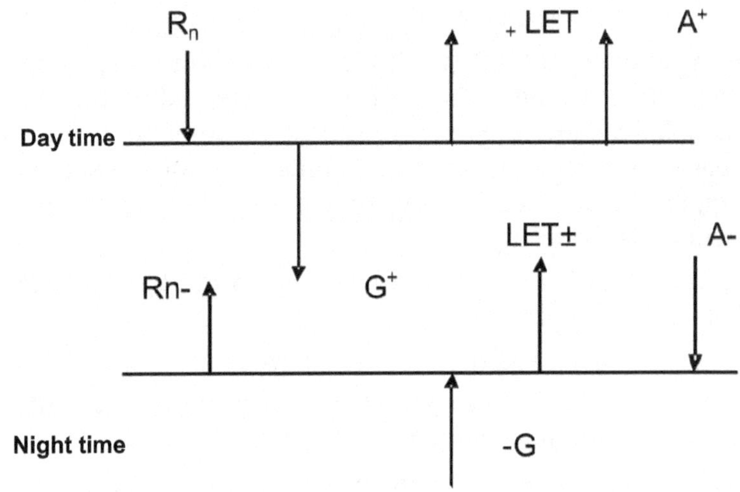

Horizontal Divergence Flux (Advection)

The available net energy over the land surface varies due to variation in surfaces, vegetation types topographical conditions.

Variations in reflectivity albedo, surface temperature irrigation water availability, crop cultivations, cultural practices, will influence the variations in net radiation values over the surfaces. This difference of net energy availability will lead to the movement of horizontal adverted energy. This additional energy will be removed from location of heat sources and will raise the net radiation available energy at locations of sink where it can be used for latent and sensible heat components for generating thermal and radiation environment for crop growth development.

Physical Storage

The total rate at which energy is stored physically (S) within the column of unit cross sectional area from soil surface to the reference height (z) of measurement of energy balance parameters as the sum of rates of change sensible, latent and vegetation heat content. Very sensitive instruments are required to quantify these small changes with time.

NET PHOTOSYNTHESIS

It is small component of net radiation energy used for net photosynthesis i.e., the net rate at which carbon dioxide is assimilated by the green vegetations per unit ground area. This rate is about 3.2 Wm^{-2} gm m^{-2} hr^{-1} of CO_2 assimilation when energy fixation for CO_2 is about 1.15 x10^4 J gm^{-1} (Baumgartner, 1965).

During daytime a maximum value ranged from 6 to 16 Wm^{-2} per 2 to 5gm m^{-2} hr^{-1} of CO_2 assimilation for vegetation species (Mointeith 1973). During night, respiration of a mature crop may reach about 1 gm m^{-2} hr^{-1} equivalent to -3 Wm^{-2}. Therefore this component of net radiation is negligible and is normally ignored in energy balance calculations.

SOIL HEAT FLUX (G)

Heat flow in the soil medium takes place through conduction phenomena. Thermal conductivity and volumetric heat capacity of soil regulate the heat flow in the medium. Fourier's laws are useful in describing the rate of heat flow through a unit cross-sectional area in soil medium called soil heat flux (G). Consider a soil column of unit cross section area with heat flux value of G(z) and G(z+ "z) at depth z and "z below the soil surface

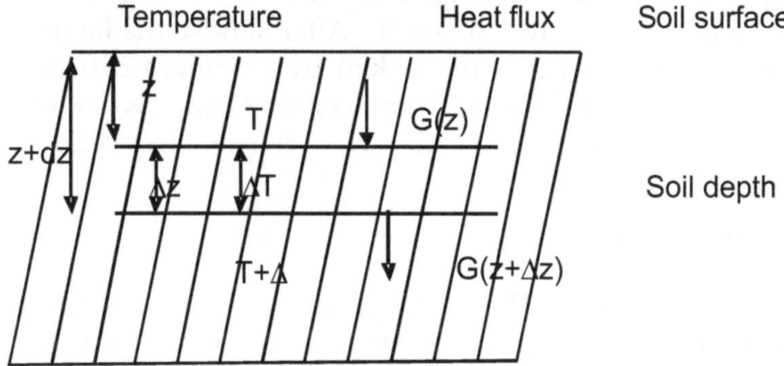

Soil temperature & heat flux at soil depth Z and z + AZ

The difference of heat flux in the layer Δz is G(z) –G(z+Δz) will raise the temperature of soil column when there is gain of flux and vice versa. Therefore downward heat flux at depth z is proportional to temperature gradient

$$G(z) = -K_s \frac{dT}{dz}$$

where, K_s is called thermal conductivity of soil, negative sign shows the direction of heat flow and temperature change is in opposite direction with depth. Now change of heat flux with depth must be equal to the rate at which heat contents in the layer Δz changes with time. The change of heat flux with depth is $G(z) - G(z+\Delta z)$.

Sensible Heat Flux (A)

Convection phenomena will allow the surface to transfer heat in terms of sensible heat depending on temperature of the surface with respect to the surrounding temperature. Using the basic flux equation the sensible heat is

$$A = -K_H \frac{d}{dz}(p_a C_p T)$$

$$A = -p_a C_p K_H \frac{d}{dz}$$

where, K_H is thermal diffusivity which is highly variable depending on the atmospheric conditions, Pa = Pressure, Cp = Specific heat of air at constant pressure 1008 J kg^{-1} $^oC^{-1}$, T = temperature ^oC in 10^1gm cm^{-1} g^{-1}. When atmosphere is stable, the energy transfer through the laminar layer above the surface having uniform temperature gradient. Eddy diffusivity takes place through molecules of air called molecular diffusivity of air (0.2×10^{-4} m^2 sec^{-1}). After sunrise the laminar layer starts diminishing as it is overtaken by turbulent diffusion and temperature gradient is not uniform over the surface. K_H now behave as turbulent diffusivity which increases with height from 0.2 to 10,000c m^2 sec^{-1}.

ENERGY PARTITION

The relative magnitude of different components of vertical energy fluxes over a horizontal surface is derived through the Bowen's ratio approach. Bowen's ratio(β) is the ratio of sensible heat flux from a surface to the latent heat flux assuming that vegetative canopy covers completely the ground surface so that soil heat flux contribution reduces to negligible component.

Therefore the available energy (H) is partitioned into

$H = A + LET = Rn \pm G$

Where, G is soil heat flux which is + ive when surface air is cooler than soil surface and –ive when soil is warmer than air near the surface,

A = Area, LET = Evaporation from vegetative surface, Rn = net radiation flux density, G = flux of soil heat by conduction per unit area.

INTERCEPTION EFFICIENCY (ξ_i)

It is the ratio of actual rate of gross photosynthesis to the maximum rate estimated for a stand of identical plants with enough leaves to intercept all the incident light (PAR). Crop canopy photosynthesis depends on the absorption and transmittance of PAR in canopy layers. Several researchers have shown that cumulative dry matter production is linearly related to the amount of intercepted radiation. Monteith (1970) provided an expression of interception efficiency as the ratio of actual photosynthesis to the maximum rate achieved by a crop canopy at full light interception expressed as

$$\xi_i = PAR_1 = \left(\frac{PAR_1}{PAR}\right) = 1 - \{s + (1-s)T\}^{LAI,}$$

Where, T is 0.07 for fully turgid leaves on clear days. PART is transmitted PAR after passing through leaf area index (LAI), S = relative sunlight area, T = resist once of medium with subscripts of M for momentum, H for heat, V for water vapour, C for CO_2 for boundary layer transfer.

Canopy Architecture

The transmission of radiation through a plant canopy depends on two features of canopy architecture. First, the concentration of green foliage matter with height influencing of the penetration of radiation beam, and secondly the orientation of leaves influencing the matual shading of the leaves.

Diffusion and Fixation of CO_2

The main parameters charactering C_3 and C_4 species are the differences in dark respiration rates (C_3 -1.67 to 5.56 kg ha^{-1}hr^{-1} and C_4- 3.34 to 10 kg ha^{-1}hr^{-1}) and maximum rate of CO_2 assimilation (C_3 - 15 to 50 kg ha^{-1}hr^{-1} and C_4 - 30 to 90 kg ha^{-1} hr^{-1}) (van Heemst, 1986).

SOLAR RADIATION AND CROPS

Crop production is exploitation of solar radiation. Quality, duration and intensity of solar radiation influences plant development and plant processes to varying degrees in different plants.

Quality

Spectral composition of solar radiation is the index of its quality. It consists of ultraviolet, visible, near infrared and infrared wave lengths of which only radiations in the range of 0.3 to 1.0μ are of importance in crop growth. Infrared radiation is associated with heat balance of plant, while that below 0.315 μ is detrimental. Near infrared range (0.76 to 0.92 μ) exerts specific elongation effect on plants, besides its importance for photoperiodism, seed germination, flowering and fruit colour development. The regions from 0.6 to 0.7 μ and 0.4 to 0.5 μ are of strong photosynthetic and formative activity, while the region from 0.5 to 0.6 μ is of weak photosynthetic and formative activity. Radiation between 0.3 and 0.4 μ produces formative effects leading to shorter plants with thick leaves.

Duration

Duration of day light period (Photoperiod) influence time of flowering of many crops. Response of plants to day light period (light) is known as **photoperiodism**. Based on flowering behaviour to the photoperiod, plants are grouped into four broad types.

1.	a.	Short -Day (SD)	:	Photoperiod less than 10 hr.
	b.	'Strict' Short - Day	:	Flowering does not occur if the day length is above or below a critical value as the case may be.
	c.	Short -day/Long day (LD/SDP)	:	Short-day plants requiring an induction period of long days.
2.	a.	Long -Day (LD)	:	Photoperiod over 14 hr.
	b.	'Facultative'	:	Effects of photoperiods additive. Long-Day plants will flower in a subsequently unfavorable photoperiod.
	c.	Long-Day/Short Day (SD/LDP)	:	Long-Day plants requiring short day induction
3.		Intermediate -Day	:	Photoperiod of 12-14 hr with inhibition of reproduction either below or above these levels.
4.		Day-Neutral (DN) length	:	Unaffected by variations in day

Photoperiodic stimulus receptors in most of the plants are mature leaves, while in few the stimulus is exerted through the growing point. In general, cold-loving plants responding to vernalization are also long-day types, while the warm-loving types are short-day types. However, some long-day plants may be induced by high temperature. Low temperature as an induction factor for short-day plants seems to be rare. This indicated that temperature and photoperiod effects be supplementary or opposing and one factor may replace the other either partly or entirely. Knowledge of photoperiodic response of crops aids in different areas of crop production.

- Production of off season crops in greenhouse culture,
- Plant breeders use the information for selecting centres of breeding if some crops and cultivars do not flower at a particular location (sugarcane, sugarbeet, onion, potato),
- Introduction of crops and cultivars in new areas, and
- Selection of proper time of sowing, especially irrigated crops for realizing higher yields.

EFFECT OF LIGHT ON PLANTS

The shorter wave lengths in the solar spectrum are harmful to the plants when exposed to excessive amounts. The atmosphere, however, absorbs almost all the shorter wave lengths. The infra-red radiation has thermal effect on plants by supplying necessary energy for evaporation of water from the plants.

The visible portion of the solar spectrum is the light with wave length ranging from 0.4 to 0.7 μ. Light is one of the important climatic factors for many vital functions of the plant. It is essential for the synthesis of the most important pigment i.e., chlorophyll. The chlorophyll absorbs the radiant energy and converts into potential energy of carbohydrates (Photosynthates). The carbohydrates, thus, formed is the connecting link between solar energy and living world. In addition, it regulates the important physiological functions like transpiration.

Effect of light on plants can be studied under four headings (i) light intensity (ii) quality of light (iii) duration of light and (iv) direction of light

1. Light Intensity : The intensity of light is measured by a standard unit called 'candle'. The amount of light received at a distance of one meter from a standard candle is known as **"Meter Candle or LUX"**.

The light intensity at one foot from a standard candle is called **"foot candle"** or 10.764 luxes and the instrument used is called as **"Lux Meter".** About one percent of the light energy is converted into biochemical energy. Very low light intensity reduces the rate of photosynthesis and may even result in the closing of the stomata which is detrimental to plants in many ways. It increases the rate of respiration. It causes rapid loss of water, while, it increases the transpiration rate of water from the plants. The most harmful effect of high intensity light is that it oxidize the cell content which is termed as **"Solarisation".** This oxidation is different from respiration and is called as **"Photo oxidation'**.

Under natural conditions, light intensity varies greatly and plants show marked response to changes of light intensities. Based on the response to light intensities the plants are classified as follows:

i) **Sciophytes:** (Shade Loving Plants) The plants that are grown better under partially shaded (low light) conditions, e.g. betel vines, buck wheat, etc.

ii) **Heliophytes:** (Sun loving plants) Many species of plants produce maximum dry matter under high light intensities when the moisture is available at the optimum level e.g., maize, sorghum, rice etc.

2. Quality of Light : Except under glass house or shaded conditions, intensity of light cannot be controlled. When a beam of white light is passed through a prism, it is dispersed into different colours with their wave length. This is called the visible part of the solar spectrum. The different colours and their wave length are as follows.

Violet & Indigo	400 - 435 nm
Blue	435 - 490 nm
Green	490 - 574 nm
Yellow	574 - 595 nm
Orange	595 - 626 nm
Red	626 - 750 nm

The principle wavelengths absorbed and used in photosynthesis are in the violet - blue and the orange -red regions. Among this, short rays beyond the violet such as x-rays and gamma rays are detrimental to plant growth. Red light is the most favourable light for growth followed by violet - blue. Ultra violet and shorter wavelengths are useful to kill the bacteria and many fungi.

3. Duration of Light: The duration of light has greater influence than the intensity. It has a considerable importance in the selection of crop varieties, the response of plants to the relative length of the day and night is known as **'photoperiodism'**. The plants are classified based on the extent of response to day length as follows.

a. Long day Plants: The plants which develops and produce normally, when photoperiod is greater than the critical minimum (greater than 12 hours) e.g., cereals, potato, sugar beet, wheat, barley, etc.

b. Short day Plants: The plants which develop normally when the photoperiod is less than the critical maximum (less than 12 hours) e.g. tobacco, soybean, millets, maize, sugarcane, etc.

c. Indeterminate or day neutral plants: Those plants which are not affected by photoperiod, e.g. Tomato, Cotton, Sweet potato, Pineapple, etc.,

The photoperiodism influences the plant characters such as floral initiation and development, bulb and rhizome production etc. If a long day plant is grown during periods of short days, the growth of internodes are shortened and flowering is delayed till the long days come in the season. Similarly, when short day plants are subjected to long day periods, there will be abnormal vegetative growth and there may not be any floral initiation.

d. Direction of Light: The direction of sunlight has a greater effect on the orientation of roots, shoots, and leaves. In temperate regions, the southern slopes show better growth of plants, than the northern slopes due to higher contribution of sunlight in the southern side.

e. Orientation of leaves: The change of position of orientation of organs of plants caused by light is usually called as **"Phototropism"**. For e.g. the leaves are oriented at right angles to incidence of light to receive maximum radiation.

Photomorphogenesis: Change in the morphology of plants due to light is called photomorphogenesis. This is mainly due to ultra-violet and violet rays of the Sun.

Instruments used for Measuring the solar Radiation are: 1. Bellani pyranometer, 2. Sunshine recorder, 3. Line quantum sensor, 4. Photometer, 5. Lux meter - measures the light intensity 6. Radiometer.

Duration of Daily Sunlight Period (Length of the Day)

The vertical rays of the Sun at noon day fall directly overhead at the equator on March 21st and this is called **"Vernal equinox"**. The vertical rays continues to move North ward to the Tropic of Cancer and are overhead these on June 21st and this date is known as **"Summer solstice"** (in Northern hemisphere). Afterwards the rays return to the equator on September 23rd and this date is known as **"autumnal equinox"**. Then it reaches the Tropic of Capricon on December 21st and this date is known as **"Winter solstice"**(in Northern hemisphere). The Summer and Winter solstices will be reverse in the southern hemisphere. At equinox, days and nights are of equal length throughout the world. In Summer solstice the day will be longer whereas in Winter solstice the day will be shorter than night. The North pole will be in day light for the full 24 hours on Summer solstice and will be dark for full 24 hours on Winter solstice of Northern hemisphere.

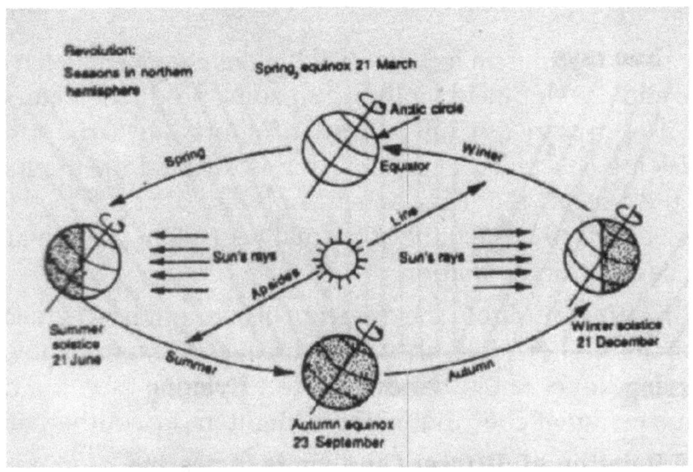

Angle of the Sun Rays

The effect of varying angle at which the Sun rays strike the Earth can be seen daily by the march of the Sun across the sky. At solar noon the intensity of insolation is the greatest but in the morning and evening hours when the Sun is at low angle, the amount of insolation is also small.

At equator the angle of incidence varies from 23 ½ ° North of the zenith to 23 ½ ° South of the zenith. The intensity of solar radiation ranges from 92 per cent on June 21st and December 21st to 100 per cent in March 21st and September 23rd. The range is only 8 per cent.

At 45 ° N latitude, the angle of incidence varies from 21 ½ ° South of Zenith to 68.5 ° South or only 21½ ° above the horizon. The variation in intensity due only to the change in the angle of incidence is from 93 per cent maximum on June 21st of 38 per cent on December 21st.

Sun rays
Morning
Noon
Evening

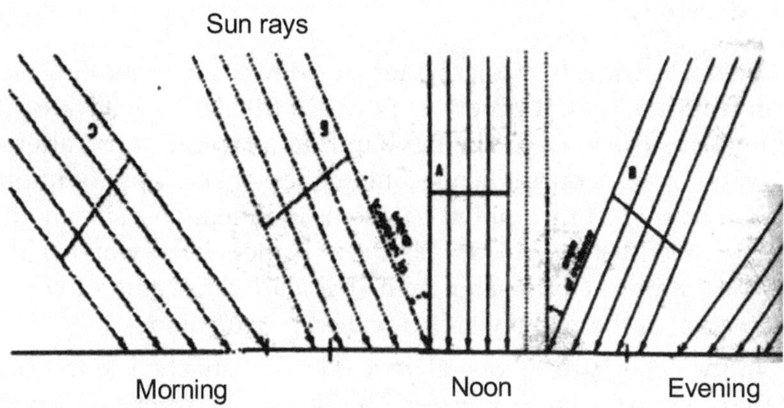

USING SOLAR RADIATION FOR GENERATION OF POWER

Most parts of India get 300 days of sunshine a year, which makes the country a very promising place for solar energy utilization. So far, photovoltaic (PV) generation has been limited to very small installations.

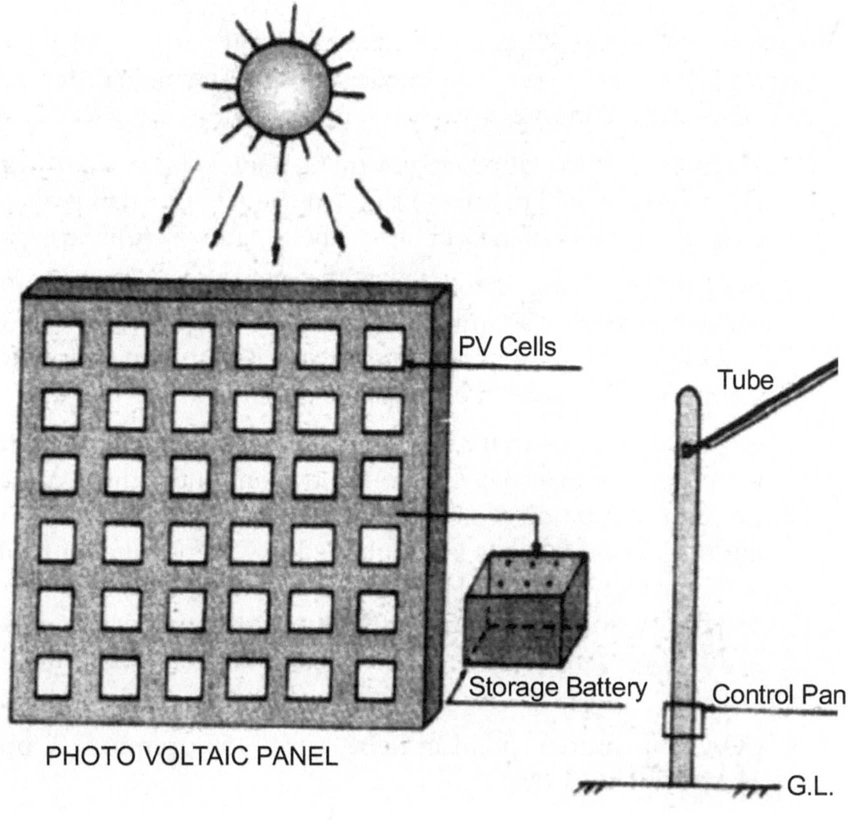

PV Cells
Tube
Storage Battery
Control Pan
PHOTO VOLTAIC PANEL
G.L.

Solar Battery

Photoelectric cell or solar battery is a device for generating electric power from sunlight. It generates power at the rate of 100 watts/m^2 of illuminated surface area. The basic unit of a typical solar battery is a thin wafer of extra pure silicon, containg less than 1 ppm of impurity. For solar battery a tiny amount of arsenic is added. This silicon-arsenic wafer is called n-type silicon. Similarly if the silicon wafer is doped with tiny amount of boron then it is called p-type silicon. The combination of these two wafers is called p-n junction.

In the solar battery, one electrical lead is attached to the surface and the other to the body of wafer. The device is a battery with positive terminal at the p-contact and the negative terminal at the n-contact. A series of such wafers (or so called solar battery cells) are used side-by-side.

USING SOLAR RADIATION BY PLANTS

The most cultivable crops are of C_3 and C_4 types by their C assimilation. The names, C_3 and C_4 refer to the length of C - skeleton of the first stable product in photosynthetic process. The main difference between these two plant types are based on following characteristics.

i. The main carboxylating enzyme in C_4 photosynthetic pathway (Dicarboxylic acid pathway) has an affinity to CO_2 that is about twice as high as that in C_3 photosynthetic pathway (Calvin cycle)

ii. In C_3 the respiratory process takes place in light which results in a dependence of assimilation rate on the oxygen concentration in the ambient air, whereas this process is absent in C_4 species. Common respiratory mechanism is dark respiration.

iii. CO_2 concentration in the inter cellular space is regulated over a wide range of external CO_2 concentrations and light intensity through adaptation of stomatal aperture, the level at which the internal concentrations is maintained in C_4 types at about half of that in C_3 types (Van Heemest, 1986; Van Keulen, 1986). Size and dimension of stomata on the upper or lower leaf give rise to differences in diffusion resistance to CO_2 and photosynthetic rate.

iv. CO_2Compensation point in light is 15 - 150 ppm in C_3 crops and 0 ppm in C_4 crops.

v. Net CO_2 exchange rate at light saturation is about 20 - 50 mg dm^{-2} hr^{-1} in C_3 crops and 70 – 100 mg dm^{-1} hr^{-1} in C_4 crops.

vi. CO_2 evolution in light takes place at high rate in C_3 crops, whereas apparently none in C_4 crops.

vii. Leaf anatomy in C_3 is diffuse mesophyll, whereas in C_4 it is mesophyll compact about vascular bundles.

viii. Radiation intensity of maximum photosynthesis in C_3 is about 0.2 to 0.8 cal cm^{-2} min^{-1} and in C_4 plants it is 1 to 1.4 cal cm^{-2} min^{-1}

ix. Maximum crop growth rate is 20 to 40 g m^{-2} $days^{-1}$ in C_3 crop, whereas in C_4 species it is 30 -60 g m^{-2} day^{-1}

Impact of Environmental Factor

The plant species differ widely to the dependency of light on the CO_2 fixation rate. Leaves of C_4 species (sugarcane, maize) provides a linear increase in CO_2 uptake with increasing levels of irradiance, whereas in C_3species leaves of rice, (grasses), beans become saturated at lower levels of irradiance Dark respiration in temperature dependent, the rate of photosynthesis and respiration increases with increase in temperature and attains a maxima corresponding to the optimum temperatures and then decreases at higher temperatures where enzymes become ineffective under normal conditions of CO_2 availability and higher saturation. The threshold temperature levels are 5 to7 ^{0} C for tropical plants, whereas for temperate and cold region plants is even below 0^0 C. The optimum temperature for agricultural $C_{4,}C_{3,}$ shade species and cold region plants are 30 to 35^0 C, 15 to 20^0 C temperate and 25 to 30 for tropical; 10 to 20^0 C, and 5 to 15^0 C respectively (Larcher, 1980). The uptake of carbon dioxide cease at 50 to 60^0 C for $C_{4,}$ and at 40 to 50^0 C for $C_{3,}$ species, soil and root respiration would also increae the release of CO_2 with increase in temperature. Even the stomatal or mesophyll resistances are influenced by temperature to alter the diffusion of CO_2.

The photosynthetic rate increases linearly with increase in CO_2 concentration in the atmosphere which have crossed 400 ppm level. In $C_{3,}$ species increase of CO_2 levels will lower photorespiration due to competition between oxygen and CO_2 molecules for the availability of free enzyme sites in the leaves.

Availability of water or soil moisture is essential for photosynthesis. Inadequacy of soil water or dryness in atmosphere reduces the photosynthesis rate through closing of stomata, decrease of optimum synthetic rates increase of resistance to diffusion of CO_2 into the leaves.

Common Questions and Answers

Why the TV programs telecast by dish tv are disrupted or blocked during rainy period?

Weakening of satellite signal during bad weather is called 'rain fade' or 'rain attenuation'. Rain fade occurs due to the presence of moisture in the air between the transmitting satellite and the receiver site. Moisture interferes with the satellite signal. The raindrops weaken the transmission by absorbing and scattering the electromagnetic signals. More moisture in the air will interfere the TV program more. Several frequencies are used to carry satellite transmissions. Earlier satellite television was broadcast in C-band-radio in the 3.4- gigahertz (GHz) to 7-GHz frequency range. Currently, digital satellite TV is transmitted in the Ku frequency range i.e. 10 GHz to 14 GHz. Lower frequencies (longer wavelengths) move through the moisture in the air better than higher frequencies (shorter wavelengths). The longer wavelengths of C-band are less susceptible to rain attenuation than the shorter Ku wavelengths. Higher frequencies experience most deterioration in their signals due to rain fade. This interference cannot be avoided completely, but its effects can be minimized and the reception of signals can be improved. One way is to increase the antennae size. But under heavy rain conditions increase in antennae size will not have any impact.

How the sun rays helps to produce vitamin-d in human's skin?

Vitamin D, a fat soluble vitamin resembles sterols in structure and functions like a hormone. Ergocalciferol and cholecalciferol, both sterols are the precursor substances for the synthesis of vitamin D and are referred to as provitamins. Ergocalciferol is designated as vitamin D_2 and cholecalciferol also as vitamin D_3. Ergocalciferol is the provitamin found in plants and cholecalcferol is the provitamin present in animals. In man the provitamin cholecalciferol (also called calciol) is synthesized from 7-dehyrocholesterol an intermediate product of cholesterol biosynthesis. The conversion of 7-dehyroholesterol into provitamin cholecalciferol takes place in the skin (dermis and epidermis) on exposure to sunlight. The provitamins D_2 and D_3 as such are not biologically active. They are metabolized identically in the human body and converted into active forms of vitamin D. The active vitamin D is calcitriol. Conversion of provitamins into active calcitriol takes place in two steps. The first step occurs in the liver where the provitamin is converted into hydroxycholecalciferol and in the second step, which occurs in the adrenal cortex of our kidney, the hydroxycholecalciferol is converted into active vitamin called calcitriol. Anyhow, for the synthesis of provitamins from which active vitamin D is formed, sunlight is very essential. The incident UV-rays in the sunlight helps for this. Humans make 90 per cent of their vitamin D naturally from sunlight exposure to the skin - specifically, from ultraviolet B exposure to the skin, which naturally initiates the conversion of cholesterol in the skin to vitamin D_3. Few foods naturally contain or are fortified with supplemental vitamin D. It is just 10 per cent of what the most we need daily. In contrast, sun exposure to the skin makes thousands of units of vitamin D naturally in a relatively short period of time. Therefore vitamin D is regarded as Sun Shine Vitamin.

Why do earthen pots lose their efficiency in keeping the water cool after being used for a few years?

Earthen pots are considered as a local fridge for keeping the water cool during summer period. As soon as the period of keeping the water is increased, the efficiency of the pot is deteriorated in cooling of water. This is due to the clogging of micro pores which are present in the pot container. Before saying this abruptly, we have to know the concept of cooling by the earthen pots.

The earthen pot acts as a medium between the outer air space and water contained in the pot. The earthen pot is made up of soil which has microniced coarse particles of soil with humus. The raw pot is burned in the furnace/kiln(sooly) for burning of humus in the soil. The well burnt pot only will give metallic sound when beaten the pot with your indicator finger back. The partially burnt or unburnt pot will not produce the metallic sound. This pot is not efficiently store and keeps the water cool. Because, the micro capillary pores are closed in this pot. The wall of the well burnt pot has two surfaces, one inner surface which is in contact with stored water and another outer surface, exposed to the outer environment. When water is stored in the pot, the water will ooze out through the micro pores present in the wall of the pot. The outside temperature of the pot is high and water vapour is low. To reduce the temperature and filling the microclimate of the pot with moisture, the ooze out water through micro capillary pores of the pot will evaporate with the help of getting temperature from the stored water in the pot.

Due to latent heat of vaporization, the heat from water in the pot is reduced. Automatically the temperature of the stored water will cool. This is continuous process up to the pores clog. When the same pot is used for few years, the micro pores will clog by microniced dust and solvent in the water. That is why we see the white encrustation/ deposition on the outer wall of the pot when the pot is unused. That is why we purchased new pot for new season of summer for natural cool water. The micro pore clogging is reason for lowering the efficiency of the pot which is used for few years.

CHAPTER - 5

Atmospheric Pressure

The atmospheric pressure is weight of the air which lies vertically above an unit area centered at a point. Normal pressure of air is 1.034/cm^2.

Pressure is defined as the force per unit area.

$$\text{Pressure} = \frac{\text{Force}}{\text{Area}}$$

At the surface of the earth the atmosphere air exerts a pressure of about 10^5 Newtons/m^2 or 10^5 Pascals, which is called one bar or one atmosphere. Evangelista Torricelli (1608-1647) devised a method for measuring the atmospheric pressure by mercury (Hg) barameter in 1643.

The approximate variation in pressure with altitude in middle latitude is as follows.

Altitude (Km)	Pressure (hPa)
100	0.0003
80	0.0104
50	0.798
40	2.87
30	12.0
20	55.3
10	265
505	500
0	1000
-50m	1020
-150m	1030
-300m	1050
-550m	1080

MSL pressure = 1013.25 h Pa

On an average at MSL, Nitrogen exerts pressure of about 760 h Pa, oxygen exerts about 240 h Pa and water vapour about 10 h Pa.

Ordinarily we fail to perceive and small change in pressure, but they are very important in bringing about changes in our day-to-day weather. Besides, air pressure is more important in weather forecasting. As climatic controls, pressure and wind plays the most dominant role through their effects on temperature and precipitation. Pressure is mainly operated by temperature and moisture in the air.

TYPES OF PRESSURE AND THEIR ORIGIN

Pressure system greatly differs in both size and duration.

There are two types of pressure system .1. High and 2. Low pressure system.

Low Pressure

The centres of low pressures are designated as depressions.

Cyclones or Lows

Prolonged low pressure points are called **throughs**. The equatorial belt or low pressure is called **doldrums** (5 ^{o} N and 5 ^{o} S to the equator) and is because of (i) Sun fall vertically all round the year making too hot (ii) water vapourisation is high and (iii) rising of air. The doldrums belt is spread on Amazon, Congo basin, Guinea belt etc.

High Pressure

The centres of high pressure are called **anticyclones or highs**. An elongated high pressure is called **ridge**. Near 30° N and S the pressure is always high due to (i) intensive hot air from the equator descends down in this belt (ii) polar air from the sub polar belts also descends here.

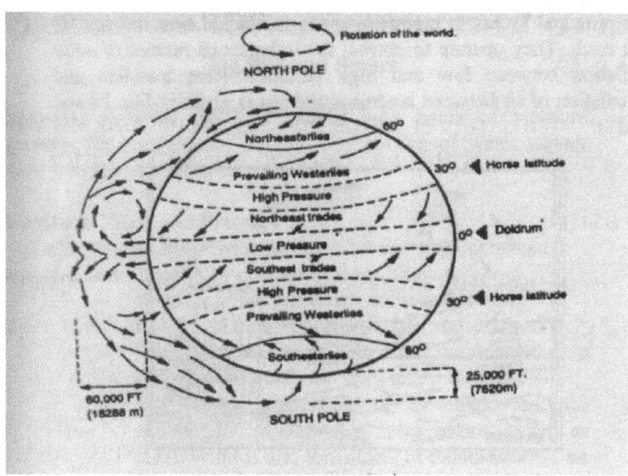

DIURNAL VARIATION OF PRESSURE

At any given place the atmosphere will be ever changing which may be periodic (regular) or non-periodic (irregular). Periodic variations are regular but they have different periodicity. Of all periodicity 12 hours (1/2 day) is of most important. This is called semi-diurnal variations of pressure. As the earth rotates, day and night are formed. Heating of atmosphere during day (due to insolation) and cooling at night (nocturnal cooling) causes rhythmic expansion and contraction, which results in pressure oscillations. Further it is assumed that atmosphere itself has 12 hour periodic oscillation. This is amplified by the day and night temperature variations. Because of this resonance, amplitude oscillations magnify (increase). Therefore a double sinusoidal pressure wave travels around the earth along with the sun. It is seen that a pressure maxima occurs at about 1000 and 2200 hrs local mean time and a pressure minima at about 0400 hrs and 1600 hrs (LMT).

In tropical regions diurnal variation of pressure is clearly observed. However it is not so clear on higher latitudes. Thus periodic variations of pressure are very complex and they are not perfectly symmetrical. These changes vary from place to place and have little influence on other meteorological parameters.

Non-periodic variations of atmospheric pressure are mostly associated with the passages of pressure systems such as Low, Depression, Cyclone. Abrupt variations of pressure are associated with thunderstorms, tornadoes, pressure shockwaves, sudden explosions etc.

FACTORS CONTROLLING PRESSURE OF AIR / CAUSES OF VARIATION

Pressure of air never remains constant and changes due to temperature, altitude, water vapour content and rotation of Earth.

1. Temperature

Hot air expands and exerts the low pressures at equator.

Cold air contracts and gives high pressure at Polar region.

2. Altitude

1. Coastal areas: at sea level, the air column exerts its full pressure.
2. Hills (we leave certain portion of the air - low pressure. For every 1000 feet of ascent, pressure drops by 1" of (one inch) Hg column (10m - 1mb).

   The relationships between altitude and temperature and air pressure are given in Fig.

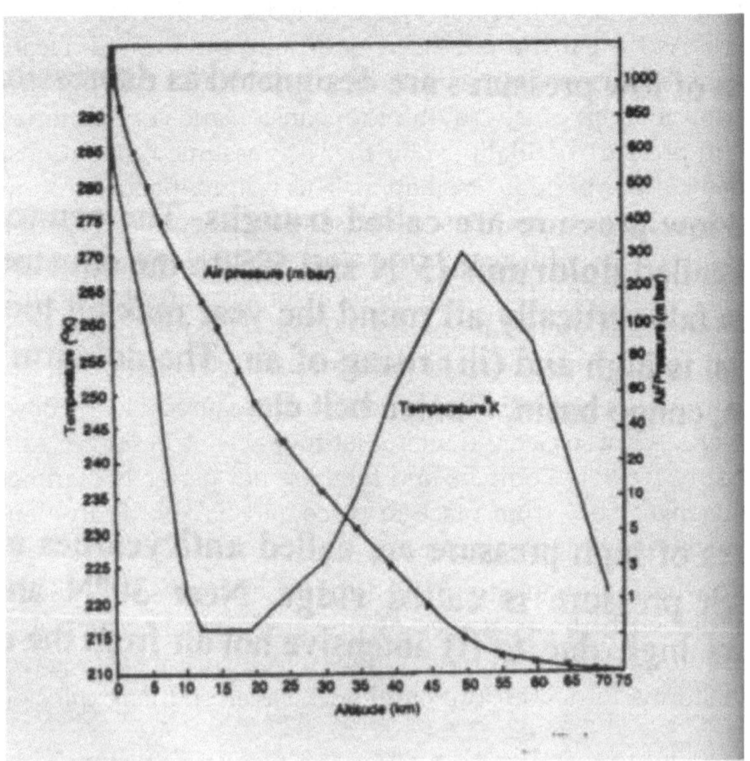

3. Water Vapour

- Water vapour content is higher in cold air than in hot air.
- Moist air low temperature gives high pressure.
- Moist air of high temperature, responsible for low pressure.

4. Rotation of Earth

By rotation at 60 - 65° N and S, pressure becomes low. Due to rotation, the air escapes and moves towards Horse latitude 30 - 35° N and S. These latitudes absorb the air from sub-polar belt, making the pressure high.

SEASONAL VARIATIONS

Pressure changes according to the season. Season changes according to the temperature, related to position of Sun.

When the Sun moves to tropic of Cancer (N) (Summer) pressure belts move to the North by 5° from their normal. Similarly in Southern hemisphere happened as Northern hemisphere. This is known as **"Sowing of the pressure belts"**

In Summer - seas and oceans have high pressure and land have Low pressure, wind moves from sea to land (high pressure to low pressure). - **"Sea breeze"**

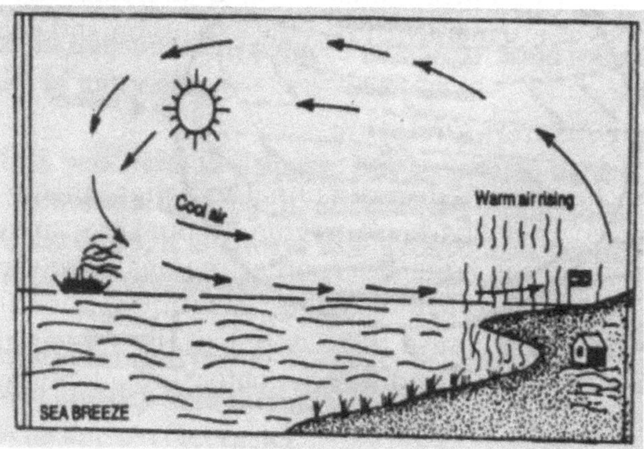

Sea Breeze

In winter - land masses have high pressure, low pressure is in sea. Wind movement from land to sea is called **"land breeze"**.

Pressure variations resulted in diurnal and seasonal variations.

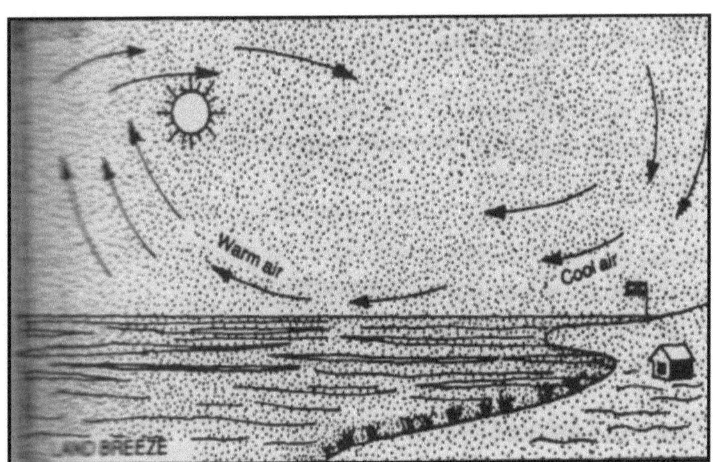

Land Breeze

DIURNAL VARIATIONS

To find out the mean daily change in air pressure, the average of hourly observed pressure for a long period of time is calculated. The mean value of the daily pressure is free from the temporary effect of atmospheric disturbances. There is a definite rhythm in the rise and fall of mercury. Insolational heating and radiational cooling are the principal reasons for diurnal variations of air pressure. In other words, pressure changes are mainly due to the expansion and contraction of the air.

Seasonal or Annual Variation

This is clearly the effect of annual variation in two amount of insolation received in a particular region. Annual pressure variation in the tropical region is larger than other regions of the world. The equatorial regions record the smallest amount of variation in their seasonal pressures, because there is practically no variation in the amount of insolation received at the equator throughout the year

High pressure - Cold season

Low pressure - Warm season

Isobars: There are lines connecting the places having the same atmospheric pressure at a given elevation. Pressure distribution charts are constructed for sea level and for number of constant pressure surfaces in the atmosphere.

Measures of atmospheric pressure: By using the instruments such as liquid containing Fortin's Barometer or without liquid Aneroid Barometer, Barograph.

Common question and answer: How does a depression bring heavy rains?

Depression : A circular low pressure area roughly 300 Km wide with a surface wind of 25 Knots (45 Kmph) is defined as a 'depression'. The circulation of air, anticlockwise in Northern hemisphere, provides sufficient updraft to generate all types of clouds and the depression could be seen like a full moon in a satellite photograph. The circular mass of clouds called 'central dense overcast' or CDO in meteorological parlance is a confirmation of the formation of a depression. With a base 5,000 feet or less, the CDO vertically extends up to 18,000 feet or so. The rains that occur, release the latent heat of condensation. A small part of it, say 10 to 20 per cent is enough to increase the circulation in the form of increased wind speeds at lower levels of the atmosphere. The depression intensifies into a deep depression. Unless the sea is cold (sea surface temperature less than 28°C) or the depression crosses a coast, this intensification keeps on going in this fashion as if there is no end to it. Around 8,000 to 10,000 feet the circulation generates a white sheet of cloud, known as altostratus (As) with a thickness that could be 5,000 to 8,000 feet. This can extend to hundreds of kilometers from the centre of the depression. This is what is seen by a common man when continuous rain of moderate intensity starts occurring over very large areas and calls it as 'depression rain' by the experience. The area covered by the rains will depend on the season. In July, when monsoon conditions are very favourable, a depression in head Bay, south of Bangladesh could bring rains of such intensity over Mumbai that it could paralyze the city. A cloud belt 200 km wide extending from Mumbai to Kolkata too will give widespread light to moderate rains. On the other hand, during the northeast monsoon period, a depression 100 Km off Karaikal and some rains over areas south of Karaikal latitude. The 'As' will not easily dissolve for one to see the sky as the moist air incursion due to the depression, will be sufficient to replenish the water vapour that is lost in the form of rains. The low clouds that surge towards the depression will partially rise up to merge with 'As' as they approach the depression and the rest will reach the depression area.

CHAPTER - 6

Atmospheric Humidity

Moisture present in the atmosphere plays a significant role in weather and climate of the region. There are three major components in the atmospheric moisture.

I. Humidity

II. Precipitation

III. Evaporation.

Humidity: The terminology related to humidity and concerned with gaseous form of water i.e. water vapour, several expressions of the amount of water vapour in the air is used. They are as follows.

a. **Absolute humidity:** It denotes the actual mass of water vapour in a given volume of air. It may be expressed as the number of grams of water vapour in a cubic meter of moist air or mass of water vapour per unit volume of air.

b. **Specific humidity:** It is defined as the moisture content of moist air as determined by the ratio of the mass of water vapour to the mass of moist air in which the mass of water vapour present.

c. **Relative humidity: (RH)** Relative humidity is common parameter for expressing water vapour content of the air. It is the percentage of water vapour present in the air in comparison with saturated condition at a given temperature and pressure. The RH can be expressed as

$$RH = \frac{100r}{rw}$$

Where, "r" is the mixing ratio of moist air at pressure and temperature and "rw" is the saturation mixing ratio at the same temperature and pressure.

i. **Mixing ratio:** The mass of water vapour per unit mass of dry air is a convenient parameter for expressing the relative composition of the mixture. It is defined as the ratio of the mass of water vapour to the mass of dry air with which the water vapour is associated.

ii. **Dew Point:** The temperature at which saturation occurs in a given mass of air. The dew point temperature is often compared with the temperature of free air and also used to predict the occurrence of fog, dew, frost or precipitation.

iii. **Vapour pressure:** This is the amount of partial pressure created by water vapour in the air expressed in the units of millibar (or) inches of mercury.

EFFECTS OF RELATIVE HUMIDITY ON PLANT GROWTH

- Increase in relative humidity decreases the temperature.
- This phenomenon increases heat load of the leaves.
- Since transpiration is reduced - not much heat energy used.
- Excessive heat due to closure of stomata, entry of CO_2 is reduced.
- Reduction in transpiration reduces the rate of food translocation and uptake of nutrients.

Very high RH is beneficial to - maize, sorghum, sugarcane (C4 Plants) harmful to sunflower, tobacco. It affect the water requirement of crops. For almost all the crops, it is always safe to have a moderate relative humidity of above 40 percentage.

MEASUREMENT OF HUMIDITY

a. Wet and dry bulb thermometer

b. Whirling psychrometer

c. Hygrometer

d. Automatic recorder

CHAPTER - 7

Wind

Air Constantly moves from place to place due to pressure differences which are caused by difference in temperature.

Wind has been defined as air in horizontal motion. According to Byers, "The wind is simply air in motion, usually measured only in its horizontal component". According to Trewaitha, "Wind is simply air moving in a direction which is essentially parallel with the Earth's surface.

There are two types of movement of wind in the atmosphere.

1. Horizontal movement
2. Vertical movement (Currents)

The primary wind circulation in Northern and Southern hemispheres is seen in figure on page 51.

The air will move vertically when it is warmer and consequently more buoyant than the surrounding air. Upward and downward air currents are referred to as updrafts and downdrafts. Particularly when considering conditions in clouds having strong vertical velocities. Vertical motions occurring in the atmosphere has great significance for the formation of clouds, precipitation, and various types of storms.

BASIC CLIMATIC FUNCTIONS OF WIND

Transporting the heat from lower of the higher latitudes, winds are the principal agents in maintaining the latitudinal heat balance by the Earth. They operate to correct the unbalanced receipt of solar radiation between low and high latitudes. Heat transfer and circulation of air between landmass and sea is given in figures in the page of 53.

Wind also provides the land masses with much of moisture necessary for precipitation through transporting of water vapour which is evaporated from ocean and land.

RELATIONSHIP OF WIND AND PRESSURE

Earth rotates from West to East along with atmosphere Atmosphere is fixed to Earth by gravitational equilibrium.

- Wind therefore moves in addition to rotation
- Horizontal motion is greater than vertical motion.
- Wind take several days to cross the ocean but up and down movement is only in few minutes.
- The vertical component of movement is much greater in small scale circulations such as thunderstorms and tornadoes. In thunderstorms, air may extent to the top of troposphere (at 6 miles) in an half to one hour.

WIND DIRECTIONS AND SPEED

- Winds are always named by the direction they come from. Wind Vane is an instrument used to find out the direction.

Lee ward refers going to direction of prevailing wind when the wind blows more frequently from one direction than from any other, it is called **prevailing wind.**

Wind direction is expressed in terms of directions in a 32 point compass and is referred by the compass point, but letter abbreviations of the directions, or by the number of degrees East or North.

The speed of the wind is measured by an instrument called anemometer. Commonly Robinson's cup anemometer is used.

Symbol	Knots	Symbol	Knots	Symbol	Knots
	Calm		23 - 27		53 - 57
	1 - 2		28 - 32		58 - 62
	3 - 7		33 - 37		63 - 67
	8 - 12		38 - 42		68 - 72
	13 - 17		43 - 47		73 - 77
	18 - 22		48 - 52		103 - 107

Symbols of wind speed.

FACTORS AFFECTING WIND MOVEMENT

a. **Barometric slope:** The wind blown towards the direction of the barometric slope i.e. from high pressure to low pressure.

b. **The isobaric gradient -** determines the steepness and gentleness of the wind

c. **Rotation of Earth -**causes the wind, to be deflected right of its path in the Northern hemisphere, but causes the wind to the deflected to the left in Southern hemisphere.

Units of wind speed: Miles per hour, Km per hour, metres per second and knots, 1 knot = 1.836 Km /hr, (or) 1.47 miles /hr (or) 0.5148 m /Sec.

Symbol of the speed of the wind is in Figure above.

EFFECTS OF WIND ON CROP PRODUCTION

Wind influences plant life, both physiologically and mechanically. The influence is more pronounced on plants on flat lands near the Sea coast and or the slopes of mountains. The wind affects the plants directly by increasing the transpiration and the intake of carbon-di-oxide and by causing several types of mechanical damage.

Physiological Effect

Under normal conditions, wind increases transpiration with increasing wind velocity, there is a greater increase in cuticular transpiration than stomatal transpiration. Wind increases turbulence in atmosphere and thus raising the supply of carbon-di-oxide to the plants and then by increasing the rate of photosynthesis. However, wind speed beyond which its rate becomes constant. When the wind is hot, it accelerates the drying of the plants by replacing humid air by dry air in the intercellular spaces. At the time of cell maturation, the dry air results in dwarfing of plants. This is because of the cells cannot attain full turgidity in the absence of optimal hydration, thus remaining at subnormal size.

Mechanical

With the influence of strong wind pressure from fixed directions, the normal form and position of the shoots are permanently deformed. Another severe injury to plants caused by strong wind is lodging. This injury is common in paddy, maize, sorghum, wheat and sugarcane. Strong winds break the twigs and shed fruits from plants. Further, crops and trees with shallow roots are often uprooted. Turbulance means violent, irregular action or swirling agitation of water, air, gas etc.

When the plant cover is not thick, strong winds remove the dry soil, exposing their roots and killing them. Eroded materials from one place become a hazard to the existing small plants in places where it is deposited. This deposited material reduces the aeration of the roots and plants. This is common in deserts, coastal areas and sand dunes.

Wind which blow from sea, do a lot of salt spraying on the coastal areas, making it impossible to grow crops which are sensitive to excessive salts.

WINDBREAKS AND SHELTERBELTS

Windbreak is a barrier for protection from winds usually associated with homestead gardens while a shelterbelt is a longer barrier than a windbreak consisting of combination of shrubs and trees intended for protecting field crops and conserving soil. Effectiveness of windbreaks depends on its density, height and length. Spacing of windbreaks depends on the density and angle of windbreaks to the direction of hazardous wind. The tallest species may be in the middle row and smallest trees or shrubs in the outer rows of windbreaks. A more or less conical cross section of windbreaks will provide the best protection from wind effect and increase the zone of protection.

The area of protection increases with the square of the length until the latter is about 24 times the height of the tree. When the wind blows at right angles to the average tree shelterbelts, wind velocity is reduced 70 to 80 per cent near the belt. In general, wind velocity is reduced by about 20per cent at a distance equal to 20 times the height of the belt. The species used for a five row belt are:

Central Rows

- *Acacia arabica*
- *Albizzia lebbek*
- *Azadirachta indica*
- *Delbergia sissoo*
- *Eucalyptus rostrata*
- *Prosopis specigera*
- *Tamarindus indica*

Flank Rows

- *Acacia senegal*
- *Anacardium occidentale*
- *Casuarina equisetifolia*
- *Cassia siamea*
- *Prosophis juliflora*

Outer Rows

- *Acacia jacquemontii*
- *Agave* sp.
- *Cassia aurticulata*
- *Glyricidia maculate*
- *Ziziphus* sp.

Structures like fences, walls, stone packing, etc. which serve as windbreaks can serve the purpose of wind erosion control. However, they may not be economical for protecting large areas.

The use of shelter belts has been one of the oldest methods employed by man for modifying the climate. Properly oriented and designed shelter belts are very effective in stabilizing agriculture in regions where strong winds causes mechanical damage and impose moisture stress on growing crops. In cold climates wind breaks, save plants from freezing and mechanical damage caused by cold winds. Wind breaks, save the loose soil from erosion, aid uniform snow cover and increase the supply of moisture to the soil in spring.

Various tall crops such as corn, sorghum, sunflower, wheat and oats are being successfully used at temporary wind barriers to protect crops like, soybean, ground nut and tomato. The primary objective of the shelter belts is to reduce the wind speed. This in turn is responsible for the amelioration of the microclimate of fields and thus increases the production.

Shelter belts: Growing row of trees (10 row plantation) to check the wind velocity at perpendicular direction to the prevailing wind.

Wind breaks: Line of trees or fence which gives protection against wind - Give good result in checking the wind velocity.

UTILIZATION OF WIND POWER

India has abundant wind resources to harness for power generation. Strong seasonal winds blow across the Indian subcontinent from April to September. The Ministry of Non-Conventional Energy Sources has estimated that the gross wind power potential of India is about 45,000 M We and has identified more than 200 sites suitable for wind power facilities. Southern India, in particular, the states of Gujarat, Andhra Pradesh, Tamil Nadu, Karnataka, Kerala, Madhya Pradesh, Maharastra and Rajasthan have excellent wind resources for conversion to electrical energy.

India ranks fifth place in the world in the number of wind power installations. Wind power installed capacity is now more than 1,700 M We, with almost all of the capacity located in the southern half of the country. India's wind power generation is mostly not yet making use of the most modern large scale technology. Wind power farms in India are made up of numerous small units, each generating in effect, only an incremental amount of electricity.

Wind blows out fire, at the same time it spreads fire. Why this phenomenon?

1: Firstly, consider the presence of a low velocity wind. (say, virtually, still air). At any moment of time this wind is just sufficient to sustain the fire. That is, it functions just as an 'element'. Obviously, it cannot provide any motive force (kinetic energy) for the fire (flame). Hence, the fire cannot spread. Ultimately, it blows out. Secondly, consider the presence of a high velocity wind. At any moment of time, there is not only sufficient supply of air to sustain the fire but also there is adequate air to provide motive force. Thus, its function is two-fold: as an element and as a motive force. Hence, the fire spreads.

2: When we blow out a small candle we are trying to push the hot flame away from the fuel and the heat is removed away from the fuel and thus reaction of burning gets stopped. For bigger fires the flame is bigger and more air is to be pushed quickly in order to push the fire away from the fuel. But when air is blown gently on small flame, fresh supply of oxygen is pushed to it for the fuel to burn and the flame may also reach new fuel because of the blowing air that would normally be out of reach. So depending on the size of the fire and how hard the air is blown the fire becomes worse or better. Thus if the fire is small enough blowing air will put off the fire. For bigger fires blowing air may fan the flames adding more oxygen and the fire will grow and the flame may reach new fuels nearby causing these fuels also to burn.

CHAPTER - 8

Clouds

Cloud has been defined as a visible aggregation of minute water droplets and / or ice particles in the air, usually above the ground level.

CLASSIFICATION OF CLOUDS

Though confusion apparently arises from the number of kinds or species, the genera seems reasonably clear out, if we are able to recognize their main characteristics.

Clouds are usually classified according to their height and appearance. For convenience, we list them in descending order. High clouds, middle clouds and low clouds are some forms do not fit in any of these categories. But fortunately their particular characteristics make them easily identifiable as vertical development clouds. We must exercise some caution in studying on height. There is some seasonal as well as latitudinal variation and there is some overlapping from time to time. However, the appearance of clouds is quite distinctive for each height category.

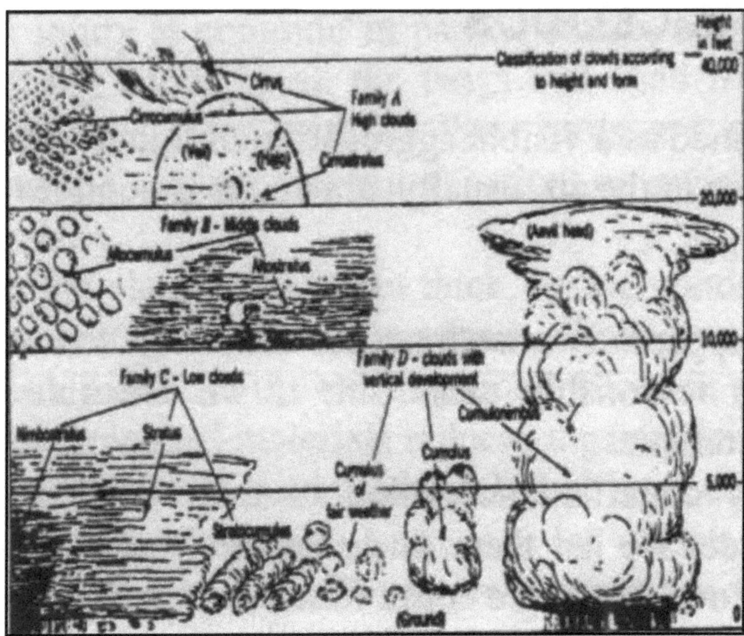

Ten main cloud genera are defined and described in the International cloud Atlas of the WMO. They can be listed according to their heights as under.

A. High (mean heights 7 to 13 km)

   i) Cirrus (Ci)

   ii) Cirro -strattus (cs)

   iii) Cirro Cumulus (cc)

B. Middle (mean heights 2 to 7 km)

   i) Alto-stratus (As),

   ii) Alto-cumulus (Ac)

C. Low (mean heights 0 to 2 km)

   i) Nimbostratus (Ns)

   ii) Strato - Cumulus (SC)

   iii) Stratus Clouds with vertical development (St).

   iv) Cumulus (Cu)

   v) Cumulo - nimbus (Cn)

1. Cirrus

Detached clouds are in the form of white, delicate filaments or white or mostly white patches or narrow bands. These clouds have a fibrous (hair like) appearance or a delicate silky appearance or both. All the cirrus or cirro-type clouds do not give precipitation.

2. Cirro - Stratus

"Transparent, whitish cloud veil of fibrous (hair like) or smooth appearance, totally or partly covering the sky and generally, producing halo phenomena". This type of cloud is so thin that it gives the sky a milky appearance.

3. Cirro - Cumulus

"Thin, White flakes, sheet or layer of cloud are without shading, composed of very small elements in the form of grains, ripples etc. This type of cloud is not common and is often connected with cirrus or cirrostratus. When arranged uniformly, it forms a "Mackerel sky". (like fish with greenish blue stripped back silvery white belly).

4. Alto - Stratus

A uniform sheet cloud of 'greyish or bluish cloud frequently showing a fibrous appearance, totally or partly covering the sky, and having parts thin enough to reveal the Sun at-least wave like as through ground glass. Alto-stratus does not show halo phenomena. This type of clouds may cover all or large portions of the sky. Precipitation may fall either as fine drizzle or snow.

5. Alto - Cumulus

"White or grey, or both white and grey, patch, sheet or layer of cloud. They have devel (heavy or large) shedding on their undersurfaces. Sometimes referred to as "Sheep clouds, Woolpack clouds".

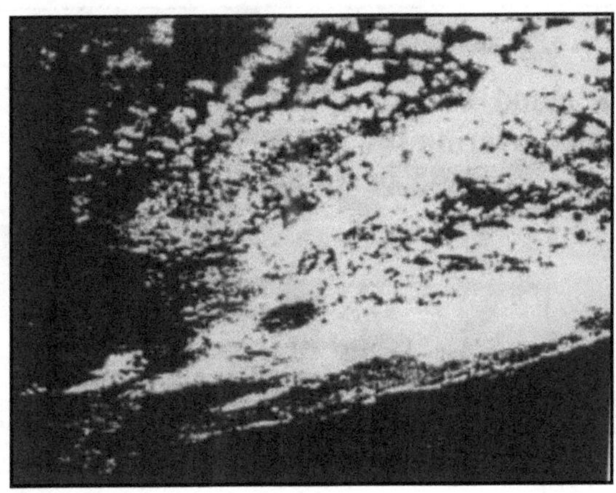

6. Nimbo - Stratus

"Grey cloud layer, often dark, the appearance of which is rendered diffuse by more or less continuously falling rain or snow which in most cases reaches the ground. It is thick enough throughout to black out the Sun. It is a rain, snow or sleet cloud. It is never accompanied by lightning, thunder or hail. Streaks of water (rain) or snow falling from these clouds but not reaching the ground are called "**virga**" (wisps or streaks of water or ice particles falling from base of a cloud but evaporating completely before reaching the ground).

7. Strato - Cumulus

"Grey or whitish or both grey and whitish patch, sheet or layer of cloud which almost always has dark parts, composed of tessellations, rounded masses, rolls etc.

8. Stratus

Generally grey cloud layer with a fairly uniform base, which may give drizzle ice prisms or snow grains, sky may be completely covered by this type of cloud. Sun is visible through thin cloud.

9. Cumulus

"Detached clouds, generally dense and with sharp outlines, develop vertically in the form of rising mounds, domes or towers, of which the bulging upper parts often resembles a Cauliflower. Cumulus is generally found in the day time over land areas. They dissipate at night. They produce only light precipitation.

10. Cumulonimbus

"Heavy and dense clouds, with a considerable vertical extent will show in the form of mountain or huge towers. This type of cloud is associated with heavy rainfall, thunder, lightning, hail or tornadoes. This type of cloud is easily recognized by the fall of a real shower and sudden darkening of the sky.

CLOUD FORMATION

Air contains moisture - this is extremely important to the formation of clouds. Clouds are formed around microscopic particles such as dust, smoke, salt crystals and other materials that are present in the atmosphere. These materials are called **"cloud condensation nucleus" (CCN).** Without these, no cloud formation will take place. Certain special type known as **"ice nucleus"** on which cloud droplets freeze or ice crystals form directly from water vapour. Generally condensation nuclei are present in plenty in air. But there is scarcity for special ice forming nuclei. Generally clouds are made up billions of these tiny water droplets or ice crystals or combination of both. These are two rain forming processes viz.1. warm 2. cold

WARM RAINS

It refers to rainfall process in the tropics. Rain occurs when the temperature is above 0° C, never colder than 0° C. When larger droplets collide and absorb smaller cloud droplets. They grow larger and larger and become raindrops. This process is known as "Coalescence."

COLD RAINS

It occurs when the cloud temperature is colder than 0° C. Clouds are usually with ice crystals and liquid water droplets. Theses crystals grow rapidly drawing moisture from the surrounding cloud droplets until their weight causes them to fall. Falling ice crystals may melt and join with smaller liquid cloud droplets. If crystals do not melt, they may grow into large snowflakes and reach the ground as snow.

CONDITIONS FAVOURABLE FOR THE OCCURANCE OF PRECIPITATION

- The cloud dimension (vertical - 7 km, horizontal 60-70km)
- The life time of the cloud (at least 2-3 hrs)
- The size and concentration of cloud droplets and ice particles.
- RH should be 75 percentage
- Wind velocity 20 kmph
- Cloud should be either cumulus or cumulo nimbus.

CLOUD SEEDING

It is the process by which the conditions of the cloud, (dimension, life time and size) are modified by supplying them with suitable nuclei at proper time and place. For accelerating the warm rain process, seeding with very large nuclei such as salt crystals can be used. In the case of cold rain process, seeding with ice nuclei such as silver iodide are used to make rain.

CHAPTER - 9

Hydrological Cycle

Hydrological cycle involves four major steps, Viz., Evaporation, Transpiration, Condensation and Precipitation. Though the cycle has neither a beginning nor an end, the concept of cycle begins with the water of oceans, since it occur nearly ¾ of the Earth's surface. Radiation from the Sun, evaporates water as water vapour from the oceans into the atmosphere. The water vapour rises and collects to form clouds. Under certain condition, the cloud moisture condenses and falls back to the Earth as rain, snow, hail etc. Precipitation reaching the Earth surface may be intercepted by vegetation or enter into the soil, may flow as run-off or may evaporate.

Evaporation may be from the surface of the ground or from free water surface. **Transpiration** may be from plants.

Condensation: The physical process by which a vapour becomes a liquid or solid - opposite of evaporation.

PRECIPITATION

Precipitation has been defined as water in liquid or solid forms falling to the Earth. Precipitation occurs in a variety of forms such as rainfall, snow, hail. Fog and dew. Fog, dew and frost are condensation forms and are not considered as precipitation.

Common precipitation forms are drizzle, rain, snow, hail and sleet etc. Precipitated moisture falling on the ground, takes various forms which depend on the following conditions: (a) the temperature at which condensation takes place, (b) The conditions encountered as the particles pass through the air, (c) the type of clouds and their heights from the ground (d) the processes generating precipitation. All forms of precipitation regardless of appearance are collectively termed as **"hydrometeors".**

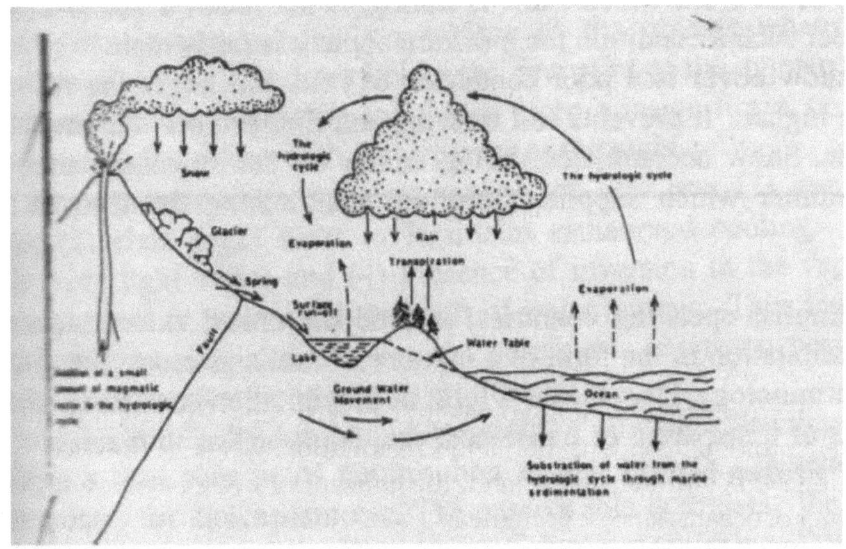

Rain: It is "precipitation of liquid water particles, either in the form of drops of more than 0.8 mm diameter or in the form of smaller widely scattered drops.

Drizzle: It is "fairly uniform precipitation composed exclusively of fine drops of water (diameter less than 0.8mm) very close to one another'. In some places drizzle is called **mist.** If the droplets in a drizzle completely evaporate before reaching the ground, the condition is referred to as mist. However, in the International codes for weather reports, the term 'mist' is used when the hydrometer - mist or fog - reduces the horizontal visibility at the Earth's surface do not less than one km.

SNOW

It is "precipitation of white and opaque grains of ice. In Winter, when temperature are below freezing in the whole of atmosphere, the ice crystals falling from the alto - stratus" do not melt and reach the ground as snow. Heaviest snowfall is reported to occur when the temperature of air from which snow is falling, is not much below 0° C, because under such a condition the moisture content is fairly high.

A snow cover is a poor conductor of heat and keeps the soil temperature higher. It prevents soil freezing and thus protects the roots of the plants. Snow accumulated during Winter on the mountains and melts in Summer which supplies water for maintaining the flow in rivers.

SLEET

In English-speaking countries, outside the united states, sleet refers to precipitation in the form of a mixture of rain and snow. But in American terminology, sleet means a form of precipitation consisting of small pellets of transparent or translucent ice, 5mm or less in diameter is called as **"Frozen rain".**

Sleet: Transparent, globular, solid grains of ice freezing of rain drops.

Hail: It is "precipitation on small balls or pieces of ice (hailstorms) with a diameter ranging from 5 to 50 mm or sometimes more, falling either separately or agglomerated into irregular lumps. It is always produced by convective clouds, usually cumulonimbus.

Hail: Precipitation in the form of balls or irregular lumps of ice.

CONDENSATION (DEW, FROST, FOG)

1) Dew is common form of condensation in the environment.
   - Dew forms on the ground and on solid objects before condensation occurs in the air.
   - Because the ground cools rapidly than the air.
   - Same way, air coming in contact with cold surface may be cooled below its dew point and gets condensed and deposited.

DEW POINTS

Temperature by which air must be cooled, at constant pressure and moisture content for saturation to occur.

Dew forms on automobiles and other metal objects first, metals cools rapidly than soil (or) vegetation.

Dew can form at any temperature above freezing point in tropics, often it forms at as high temperature as 21°C.

On an average atmosphere contains water vapour in varying degrees from 0 to 3 per cent. On clear nights, this moisture combusts as small water droplets known as dew on surface of the objects when the temperature of the surface drops below the dew point of the ambient air temperature. Cooling inside the soil during night is insignificant.

Important conditions for formation of dews are (1) enough moisture (2) clear night skies to facilitate cooling due to radiation. (3) Calm or very light winds and (4) Presence of inversion in the vapour pressure gradient or downward transport of water vapour. Thus in spite of high humidity, dew deposits are not possible in monsoon because radiational cooling process is greatly retarded.

In the arid to dry sub-humid spectrum of climatic classification, dew plays a dual role in its contribution to plant growth. It actively provides water for direct plant use. The passive role is to delay the rise in temperature to lethal levels on the following morning thereby reducing evapotranspiration.

Rabi crops in India are sown during Oct-Nov and harvested during Mar-April. These crops season falls outside the main rainy season, with very little irrigation facilities, the crops depends heavily on dew deposits for their sustenance and growth.

Dew deposits vary from season to season and also with heights because of variations in the sensible heat transfer (Effective outgoing radiation) within the plant community dew falls vary according to prevailing weather and surface conditions.

y (dew amount in mm) = ax^b, a & b = constant 2.95 & 0.07 respectively x = height (cm). Ramdas (1943) believed that dew deposits in India were more at 30 cm height above ground. Raman et al.(1973) found that dew deposits at 50 cm height were consistently higher than 25 cm height. Chowdhury *et al.*(1990) found that largest dew deposits at 100cm height. The deposit at this height was nearly 25 per cent more than that at 5 cm, 10 per cent more than that at 25cm and about 5 per cent more than at 50 cm.

Maximum dew amounts:

Country	Max. dew in a night (mm)	Reference
Australia	0.1 - 0.3	Tuller and Chilton (1973)
British Colombia	0.15 -0.2	Muller (1968)
England	0.2 - 0.5	Monteith (1969)
Germany	0.3 - 0.4	Hofmann (1958)
India	0.1 - 0.3	Raman et.al (1973)
Isreal	0.2	Duvdevani (1957)

Maximum dew accumulation is conditioned by large scale synoptic features. Local factors like orography, Proximity to large water bodies, distance from the forest edge, cause changes in the net outgoing radiation, air temperature and relative humidity and produce variation in dew falls.

Dew falls (Total amount for the month) DD/RF

Rainfall (Total amount for the month) DD/PE

Potential Evaporation (Total Amount for the month)

The ground surface creates favourable soil moisture condition for dew formation. This is because that it is not only cools the atmosphere but also provides enough moisture for condensation.

The dew falls, potential evaporation and dew durations for the rabi, winter, upto begining of summer months are given for the place of Adenthurai, Tamil Nadu where Paddy cultivation is predominant.

Aduthurai (11^0 01 N, 79^0 32′ E)

Month	DD (mm)	RF (mm)	DD/RF	PE(mm)	DD/PE	Dew duration (min)
October	1.547	232.2	0.01	131	0.0012	25
November	2.286	386.6	0.04	111	0.0030	52
December	2.840	1844	0.01	118	0.0024	56
January	5.238	76.2	0.07	132	0.0039	55
February	5.346	21.3	0.25	137	0.0039	48
March	5.130	19.1	0.27	172	0.0030	38

Frost: Frost is a form of condensation that occurs on solid surfaces when the dew point is below freezing point (0° C). Water vapour sublimites as a solid on solid. (passing directly from a gas to solid). Frost can form at a point below freezing point. Occasionally, vegetation is damaged. Frost occurs mostly in low places like valleys of the mountains where there is no outlet. Conditions required for the formation of dew and frost is (a) clear sky (b) cool air and sufficient moisture to reach the dew point with a moderate amount of cooling.

Fog: It is condensed water droplets suspended in air in the lower atmosphere, surface of the Earth. Fog reduces the horizontal visibility. They frequently occur in super cold liquid at temperature much below the freezing. Accumulation of dust or smoke, fog in air is called dust fog or smoke fogs. Thick fogs are more frequent in smoky cities. The blend of smoke and fog is called 'smog'.

TYPES OF PRECIPITATION

There are three types of precipitation

1. Conventional Precipitation

Due to convective in the form of turning of moist air, heavy and showery precipitation is most likely rain, snow showers, hail and snow pellets.

2. Orographic Precipitation

Precipitation resulting from raising and cooling of air masses when they are blocked by topographic barriers (mountains). Barrier is an important factor in increasing the rainfall on windward slopes. High annual rainfall are common where the mountains act as barriers across the paths of moisture bearing winds (Chirrapunji). Indirect effect- force moist air upward, they hinder the passage of low pressure areas and also promote convention (due to differential heating along the slopes).

3. Cyclonic Precipitation

According to type of low pressure system and its stages of development

Precipitation characteristics: Air masses moves from high pressure area along with clouds. Cyclone brings rainfall in the tropical region majority due to pressure gradient.

CHAPTER - 10

Evaporation and Transpiration

The change of state of water from solid and liquid to the vapour and its diffusion into the atmosphere is referred as **Evaporation**. In the meteorology, evaporation is defined as the maximum possible loss of moisture from a wet, horizontal, flat surface exposed to weather which exist in the vicinity of plants.

FACTORS AFFECTING EVAPORATION

1. Those affecting water supply at the evaporating surface. i.e., soil and plants including soil storage capacity, rainfall and irrigation and
2. Those affecting energy supply to the evaporating surface like solar radiation.

TRANSPIRATION

Most of the water absorbed by plants is lost to the atmosphere. This loss of water from living plants is called **transpiration**. It can be stomatal, cuticular or lenticular.

FACTORS AFFECTING TRANSPIRATION

Light, humidity, temperature, wind, root/shoot ratio, availability of water to plants, leaf characteristics etc., affect the transpiration.

EVAPOTRANSPIRATION (ET)

Evapotranspiration is the summation of combined loss of water through evaporation from the soil and transpiration from the plants.

POTENTIAL EVAPOTRANPIRATION (PET)

It is defined as the amount of water which will be lost from an extensive water surface or soil completely covered with actively growing vegetation where there is abundant moisture in the soil at all times.

Evapotranspiration is also called water use (WU) or consumptive use (CU). The factors influencing ET are climate and management practices.

Evaporation

- One of the four components of the endless hydrological cycle (E-T-C-P)
- Most of the water vapour comes from ocean.
- It is also important in agriculture as it affects
  - Soil conditions
  - Plant growth - crops
  - Water storage - dams.

Evaporation Depends Upon

- temperature of the water surface
- vapour pressure of the air
- wind movement (Removes moisture)
- Salinity - presence of dissolved minerals / salts reduce evaporation, from sea it is 5 per cent less than pure water.

The pressure exerted by the water vapour in the air known as **"vapour pressure"**. Evaporation is more when there is greater pressure difference between vapour pressure and saturation vapour pressure.

Factors Which Affect ET from Plant and Soils are

A. Those affecting water supply i.e.,
   - Soil storage capacity

- Rainfall.
- Irrigation

B. Those affecting energy supply,

- Solar radiation, temperature, humidity, wind etc.,

PE - Potential evapotranspiration

Evaporation from a free water surface

AE - Actual Evaporation

AE is always less than PE.

CHAPTER - 11

Cloud Seeding

Cloud usually consists of very small droplets of water which cannot fall appreciably below the cloud base without evaporation. In this condition clouds are stable. Rain drops are about a million times heavier than cloud drops and rain develops only if the cloud droplets grow by some mechanism. Cloud Seeding mechanism depends on the difference in vapour pressure over ice and super cooled cloud, water evaporates of the droplets and migrate to the ice crystal, which grows and begins to fall increasing their growth rate by coalescence with other droplets and eventually melting and falling as rain drops. Presence of ice nuclei determines whether ice crystals form in a supercoold cloud or not. Ice crystals form on naturally occurring nuclei at temperatures of - 10^0 to 30^0 C. At warmer temperatures, either the efficient ice-forming nuclei are often deficient in the atmosphere or the cloud may occur with tops supercooled but not to an extent which activates the ice nuclei available. It should be possible to initiate precipitation by introducing ice crystals or by supplying artificial nuclei.

DRY ICE SEEDING

It is to change supercooled cloud droplet to ice crystal in the laboratory and in the atmosphere by dropping pellets of dry ice into the cloud and causing large number of ice crystals to appear spontaneously. Dry ice is solid frozen carbon dioxide at a temperature below -80^0 C. An air craft drops dry ice pellets of size 0.5 to 1.0 cm at the top of supercooled clouds. They fall through the cloud and result in a sheet of ice crystals. This method is reliable for stimulating showers from cumulus clouds provided their tops have temperatures lower than -5^{0}C and the clouds last more than 30 minutes.

SILVER IODIDE SEEDING

Minutes crystals of silver iodide prodused in the form of smoke acts as efficient ice forming nuclei at temperatures below -5^{0}C to produce enormous number of nuclei (10^{15} per gram of silver iodide). When this smoke is introduced into supercooled cloud, some ice crystals appear when temperatures fall below - 4^{0}C, but its formation rapidly increases with decreasing temperatures. Cloud seeding by silver iodide is done either from ground generators or from airborne generators. Substances other than silver iodide as artificial nuclei are lead iodide, cupric sulphide, cupric oxide, ammonium fluoride, cadmium iodide and iodine. However, all these are not as effective as silver iodide.

WARM CLOUD SEEDING

It is a process in which nuclei of droplets grow to radii of several microns in coalescence between drops. Once drops have grown to about 40 μ dia, they are capable of with and sweeping up smaller droplets in their path. Once the droplets reach the size of 60 μ dia, coalescence dominates and condensation can be virtually ignored. Coalescence process is mainly responsible for growth of rain drops in warm cloud. In this case, seeding would be based on the assumption that the presence of comparatively large water droplets is necessary to initiate the coalescence process and some clouds do not precipitate or do so inefficiently due to absence of such large water droplets. This deficiency can be remedied by introducing in the cloud, water droplets or hygroscopic nuclei. Salt particle of 3 μ becomes 50 μ particle of brine in saturated air of cloud. Water droplets of 20 to 30 μ introduced in the cloud through aircraft at the rate of 30 gallons per minute appears to be effective for obtaining the rain.

Experiments have clearly established that cloud seeding with silver iodide or dry ice in the case of cold clouds and sodium chloride or minute water droplets in the case of warm clouds can stimulate rain from suitable clouds. However, practical experience as of now, may not suggest recommending it for adoption as an option for drought during crop season.

CHAPTER - 12

Weather Aberrations (Weather Hazards and their Mitigation)

The march of weather factors during crop period is never smooth and regular. Agriculture, because of its dependance on climate is inevitably subjected to unpredictable long and short term fluctuation characteristic of atmosphere conditions. Climate introduces an element of risk in farming which makes it particularly hazardous occupation.

A Weather hazard in this context can be defined as any type of extreme weather condition that can damage the crops and animals leading to abnormal decrease in production and income. The extent to which these hazards create risk depends on the time of occurrence, intensity and duration of the hazard besides the age, stage of development and inherent resistance of the crop or livestock involved. Major abnormalities affecting the crop production are:

- Excessive rains (floods),
- Scanty rains (drought).
- Untimely rains,
- Storms, cyclones and depressions,
- Thunderstorms, hailstorms and dust storms,
- Cold waves accompanied by frost,
- Heat waves,

- Excessive or defective insolation, and
- High winds.

The question of modifying the weather abnormalities to the extent of minimizing its influence in a manner similar to that of other inputs, as yet, outside human control. However, while weather itself cannot be modified, its hazardous effects can often be mitigated by appropriate management practices.

EXCESSIVE RAINS

When heavy rains occur particularly over catchments area of rivers, the magnitude of floods may well be imagined. For such large scale phenomena, the remedies depend on large scale planning by the government for multipurpose or hydroelectric schemes. Planned afforestation of the denuded area is one of the remedies which should receive every support from the government.

Watershed development programme, under implementation during the eighth five year plan, appears to be promising in the direction of soil erosion and run off control leading to water flows into the rivers. For floods, individual farmer can only keep the drainage channels of his fields open and go for flood resistant cultivars, especially in the case of rice.

MEASURES FOR HEAVY RAINFALL SEASON

i. The national, state and local level disaster management organisations may enlist coordination, complementation and convergence of line departments, agencies and voluntary organisations.

ii. Restoring all kinds of communications including roads, bridges, telecommunication and power supply is necessary for effective and efficient response.

iii. Setting up of relief camps, shelters, arranging safe drinking water, food, clothing, medicine etc. for human beings.

iv. Setting up of cattle relief camps and making provisions for fodder, feed, water and veterinary services.

v. Sanitation in relief camps and cattle shelters will ensure the desired outputs and outcomes. Safe disposal of carcasses of livestock and wild life is called upon to avoid outburst of infection and diseases.

1. Land and Production Management

Soil erosion solutions in the upper catchments, down-stream river training works, deposition of sediments and alluvium in flood plains, livestock rearing, crops/fodder production, horticulture, aquaculture, ground water recharging and its utilization during post-floods seasons/ periods need timely attention. In spite of miseries, the floods also provide post flood opportunities to compensate production losses with intensive management of pre-*rabi* and *rabi* season crops and other production systems. Detailed contingency plan will follow.

2. Rehabilitation

Repair and reconstruction of houses, roads and bridges and other infrastructure with appropriate resources will facilitate mitigation. River training and channelization may be taken up during workable season.

3. Safety Nets

Comprehensive livelihood insurance, construction on stilts in low lying areas, creation of storage and warehousing facilities for pre-positioning of supplies, diversification in family income etc. should be part of the strategies.

DROUGHT

Low rainfall or failure of monsoon rains is a recurring feature in India. This has been responsible for droughts and famines. The word drought generally denotes scarcity of water in a region, through, aridity and drought. They are due to insufficient water. Aridity is a permanent climatic feature and is the culmination of a number of long term processes. However, drought is a temporary condition that occurs for a short period due to deficient precipitation for vegetation, river flow, water supply and human consumption. Drought is due to anomaly in atmospheric circulation.

DEFINITION OF DROUGHT

There is no universally accepted definition for drought. Early workers defined drought as prolonged period without rainfall. According to Ramdas (1960) drought is a situation when the actual seasonal rainfall is deficient by more than twice the mean deviation. American Meteorological Society defined drought as a period of abnormally dry weather sufficiently prolonged for lack of water to cause a severe hydrological imbalance in the area affected.

Prolonged deficiency of soil moisuture adversely affects the crop growth indicating incidence of agriculture drought. It is the result of imbalance between soil moisture and evapotranspiration needs of an area over a fairly long period as to cause damage to standing crops and to reduce the yields.

CLASSIFICATION OF DROUGHT

Drought can be classified based on duration and nature of users. In both the classifications, demarcation between the two is not well defined and many a time overlapping of the cause and the effect of one on the other is seen. Droughts are classified into four kinds.

1. Permanent Drought

This is characteristic of the desert climate where sparse vegetation growing is adapted to drought and agriculture is possible only by irrigation during entire crop season.

2. Seasonal Drought

This is found in climates with well defined rainy and dry seasons. Most of the arid and semiarid zones fall in this category. Duration of the crop varieties and planting dates should be such that the growing season should fall within rainy season.

3. Contingent Drought

This involves an abnormal failure of rainfall. It may occur almost any where especially in most parts of humid or sub humid climates. It is usually short, irregular and generally affects only a small area.

4. Invisible Drought

This can occur even when there is frequent rain in an area. When rainfall is inadequate to meet the evapotranspiration losses, the result is borderline water deficiency in soil resulting in less than optimum yield. This occurs usually in humid regions.

Droughts area also classified based on their relevance to the users.

a. Meteorological drought

It is defined as a condition, where the annual precipitation is less than the normal over an area for prolonged period (month, season or year).

b. Atmospheric drought

It is due to low air humidity, frequently accompanied by hot dry winds. It may occur even under conditions of adequate available soil moisture. Plants growing under favourable soil moisture regime are usually susceptible to atmospheric drought.

c. Hydrological drought

It is the result of soil moisture stress due to imbalance between available soil moisture and evapotranspiration of a crop. It is usually gradual and progressive. Plants can therefore, adjust at least partly, to the increased soil moisture stress. This situation arises as a consequence of scant precipitation or its uneven distribution both in space and time. It is also usually referred as soil drought.

Relevant definition of agricultural drought appears to be a period of dryness during the crop season, sufficiently prolonged to adversely affect the yield. The extent of yield loss depends on the crop growth stage and the degree of stress. It does not begin when the rain ceases, but actually commences only when the plant roots are not able to obtain the soil moisture rapidly enough to replace evapotranspiration losses. Important causes for agricultural drought are:

- Inadequate precipitation
- Erratic distribution
- Long dry spells in the monsoon,
- Late onset of monsoon, and
- Early withdrawal of monsoon.

PERIODICITY OF DROUGHT

The Indian Meteorological Department examined the incidence of drought for the period from 1871 to 1967, utilizing the monthly rainfall of 306 stations in the country. It was seen that during 1877, 1899 1918 and 1972 more than 40 per cent of the total area experienced drought. General observation on the periodicity of drought is given.

Meteorological subdivisions	Period of recurrence of drought
Assam	Very rare, once in 15 years
West Bengal, Madya Pradesh, Konkan, Coastal Andhra Pradesh, Kerala, Bihar, Orrisa, South interior Karnataka, Eastern Utter Pradesh, Gujarat, Vidharbha, Rajasthan	One in 5 years
Western Uttar Pradesh, Tamil Nadu, Kashmir, Rayalaseema and Telangana	Once in 3 years
Western Rajasthan	Once in 2 $_{1/2}$ years.

Frequent crop failure in dryland agriculture is due to droughts or prolonged dry spell during the crop period. Several management practices have been recommended to mitigate the adverse effect of drought.

- Increasing the water holding capacity of soils through the application of bulky organic manures,
- Soil and moisture conservation practices to minimize runoff,
- Collection of runoff water during periods of heavy rains for use during drought period,
- Minimising evaporation and transpiration losses through mulches, shelterbelts and chemicals,
- Introduction of drought resistant and drought escaping crops and cultivars,
- Farm forestry for sustainable production in dryland farming, and
- Identification of remunerative cropping systems for dryland agriculture.

UNTIMELY RAINS

Onset and withdrawal of monsoon largely determine the success of dryland agriculture. Late onset of monsoon delays sowing of crops leading to poor yields. Early withdrawal of rains affect the yield due to soil moisture stress especially when the kharif crops are at critical stages of grain development. Continuous, cyclonic rains from August lead to problem in harvesting the crops, especially groundnut in south India.

Land should be prepared for kharif crops, taking advantage of premonsoon showers, such that sowing could be completed, taking advantage of earliest monsoon rain. Alternately, short duration pulse crops are recommended for abnormal delay in monsoon. As it is not possible to predict withdrawal of monsoon, it is always better to go for intercropping long and short duration crops to avoid total crop failure, especially in areas of frequent crop failures.

STORMS, CYCLONES AND DEPRESSIONS

One of the features of colossal meteorological phenomena like cyclonic storms and depressions can be predicted and forecasted. On receipt of warnings, precautionary measures can be taken for minimizing the risk. Warning of heavy rain, especially during sowing and harvest times will be practically useful as sowings can be postponed and harvesting hastened, wherever such disturbances are expected. Weather warning on such aspects leads to efficient use of agrochemicals used for plant protection and weed control.

The recent Haiyan cyclone (November, 2013) affected the Phillipines and devasted the infrastructure. More than 10,000 people were dead. Such huge loss will occur when the precautionary measures not followed after warning given by the meteorolgical departments.

THUNDERSTORMS, HAILSTORMS AND DUST STORMS

These are comparatively local in character. They usually occur before the onset and after the withdrawal of monsoon. Though short in duration, the precipitation and associated squalls are often very violent.

Hailstorms are very destructive to standing crops and even to livestock and human life. Protection against them is difficult, particularly under rural conditions. However, farmers warned before harvest their crops if they are already ripe. Shelter belts and wind breaks can offer considerable protection to horticultural crops.

COLD WAVES AND FROST

Cold waves and frost are common during winter in NorthEast regions of India. There is considerable damage to grain and horticultural crops due to frost and freezing temperature. Many tropical and subtropical plants are killed even at temperatures higher than about 0^0C. Information on frequency and intensity of frosts can indicate frostfree growing season, which would help in selecting suitable crop species and varieties for area subjected to frost. Knowledge of critical temperatures for frost damage would be useful for planning frost prevention measures.

- For better frost protection, the soil must be moist, compact and weed free. Frost damage is more common in sandy soil than in heavy soil under similar temperature conditions,
- Straw mats, screens of dry grass netting, wax paper, plastic covers known as hot caps should be placed over small plants during late afternoon and removed early in the morning for frost protection,
- Smoking and fogging by burning wood, straw, straw dust, sump oil, tar, naphthalene, etc.,
- Creation of frost smoking in which artificial fog or chemical mist with particles sufficiently large to prevent longwave radiation.
- Use of high speed fans for temperature inversion near ground.
- Flooding the field for increasing the thermal capacity and conductivity for releasing latent heat when water freezes, and
- Replacing the heat lost through radiation by heat emitted from suitable heaters or small fires. Common materials for heating are wood, coal, charcoal, diesel oil. Heaters are the best for frost protection.

HEAT WAVES

Just as cold waves are injurious to crops in winter, heat waves are injurious in summer. Deccan and central parts of the country experience hot waves during March, April and May. During this period, temperatures go beyond 43^{0}C. Blowing of hot winds leads to pollen and premature fruit drop. Most of the water bodies dry up in a short time leading to water shortage.

Heat evasion by shading of plants appears to be effective to minimize the adverse effect of heat wave. A number of shade structures from wood or fibre are used, especially to protect vegetable crops grown on sandy soils during summer. Windbreaks also can reduce the heat wave effect by decreasing the velocity of hot winds.

EXCESSIVE / DEFECTIVE INSOLATION

During clear day in summer, soil temperatures reach as high as 70^0C, over blacksoils. High soil temperature affect seed germination and functional activity of roots. Irrigation can depress the soil temperature, but the effect lasts only as long as there is enough moisture on the surface layers. A layer of chalk or surface mulch of straw or any other organic waste helps not only to keep the soil layers cool, but also to conserve soil moisture.

When the insolation is weak, soil temperature decrease to the extent of affecting plant growth. A black cover charcoal powder is useful for absorbing insulation and heating up the soil.

HIGH WINDS

Wind affects the crops directly by increasing the evapo-transpiration and causing several types of mechanical damage including lodging. When the wind is hot, it accelerates the plant desiccation. Use of shelterbelts is the only practical remedy for protecting the crops from high winds. Shelter belts are primarily meant for altering wind speed and its direction. They also aid in soil conservation and efficient use of soil moisture by reducing evapo-transpiration rate of crops. In practice, there are three main types of shelterbelts.

1. Dense shelterbelts with little permeability,
2. Shelterbelts with medium and even permeability from soil surface to top of the belt, and
3. Alley type shelterbelts which may be rows of tall trees without shrubs as an under-storey and which are open in the lower parts.

Dense shelterbelts are impermeable to air flow and hence exhibit sharp peaks of stream lines at distances of about 1 to 2 times the height of the belt. The stream lines, however, descend steeply and become horizontal at about 15 to 25 times the height of the belt. Thus, as compared to permeable belt, larger wind reduction is achieved immediately behind the dense belt. However, the original wind velocity is restored at much shorter distance in the case of dense belt than permeable belt. Thus, the wind reduction zone is much greater in the permeable belt.

In alley type belts, as there are gaps between the trees, many steamlines pass through the lower part of the belt. The wind reduction in immediate neighbourhood, will be lower than in other types. However, the zone of wind reduction would be wide in this case also.

For the best wind speed reduction and greatest downward influence, the shelter should be more porous on lower heights. In fact, the density can increase with height in proportion to the logarithmic nature of the wind profile. The optimal degree of permeability of shelterbelt is about 30-35 per cent, if large cultivated fields are to be protected. This would mean that there should be many small gaps in the belt such that their total area works out to be 30-50 per cent of the total area of the belt.

DRY SPELL

The interval between the end of a seven day wet spell, beginning with the onset of effective monsoon and another rainy day with 5e mm of rain (where "e" is the average daily evaporation) or the commencement of another seven day rainy spell is called the first dry spell. If the duration of this dry spell exceeded certain value, depending on the crop - soil complex of the region, this dry spell was called critical dry spell.

Criteria for forecasting rainfall characteristics (like onset of effective monsoon)

1) The first day rain in the seven - day spell signifying the onset of effective monsoon, should not be less than "e" mm where "e" mm, was the average daily evaporation.
2) The total rain during seven -day spell should not be less than 5e + 10mm.
3) At least four of these seven days should have rainfall, with not less than 2.5mm of rain on each day.

WET SPELL

A wet spell is defined as a rainy day with "x" mm of rainfall or a seven day spell where the total amount of rainfall equals "x" mm or more and the condition that four out of seven days must be rainy with rainfall more than 2.5 mm on each day. In this, "x" is the amount of rainfall which brings the top 50 cm soil layer to field capacity. The water holding capacity varies with the type of soil as also affect the value of "x".

For example, the value of "x" is equal to

83 mm for light soils

125 mm for medium soil and

166 mm for heavy soils of Gangetic plains.

CRITICAL DRY SPELL (CDS)

CDS defined as the duration between the end of a wet spell and the start of another wet spell during which 50 per cent depletion of available moisture occurs in the top 50 cm soil layer.

It is calculated by

$$CDS = \frac{AMD}{ET}$$

Where, CDS in days

AMD = 50 per cent of the available soil moisture in the top 50 cm soil layer, expressed in terms of depth (mm).

ET = Average maximum daily ET of a crop (mm/day)

DRY-FARMING AREAS OF TAMIL NADU WITH ITS RESOURCE CHARACTERIZATIONS

The Semi arid tropic area facing high incidence of solar radiation, high air and soil temperatures and extremely varying rainfall. Droughts and floods are common occurrences in that area in one year or another. Deficit rainfall years may be followed by similar years with excess rainfall in uncertain pattern. In these regions, conventional dry farming is risky and farmers are reluctant to invest heavily in crop production. Hence the Dry farming or dry land farming may be defined as: "a practice of growing profitable crops without irrigation in areas which receive an annual rainfall of 500 mm or even less".

CHARACTERISTICS OF DRY LAND AGRICULTURE

Dry land areas are distinctly noted with the following features:

1. Uncertain, ill -distributed and limited annual rainfall
2. Occurrence of extensive climatic hazards like drought, flood etc.
3. Undulating soil surface
4. Occurrence of extensive and large holdings
5. Practice of extensive agriculture *i.e.* prevalence of monocropping etc.

6. Relatively large size of fields
7. Similarity in types of crops raised by almost all the farmers of a particular region
8. Very low crop yield
9. Poor market facility for the produce
10. Poor economy of the farmers and
11. Poor health of cattle as well as farmers.

IDENTIFICATION OF DRY FARMING DISTRICTS IN TAMIL NADU

Among 32 districts of Tamil Nadu, twelve have been identified pre-dominantly as rain fed farming districts, with scope of dry-land farming based on the rainfall pattern.

The dry farming districts in Tamil Nadu are given in Table.

Sl.No.	District	Sl.No.	District
1	Vellore	7	Perambalur
2	Thiruvannamalai	8	Pudukottai
3	Namakkal	9	Madurai
4	Erode	10	Ramnad
5	Trichy	11	Sivagangai
6	Karur	12	Thoothkudi

RESOURCE CHARACTERIZATION IN DRY FARMING DISTRICTS

Rainfall-Pattern and Length of Growing Period

According to characteristics of dry farming, either there will be no rain at all or there will be torrential rain with very high intensity. Thus, in the former case the crops will have to suffer by a severe drought and in the latter case they suffer either flood or water logging and they will be spoil. In case of very heavy drownpour, the excess water gets lost as run-off which goes to the ponds or down drawndened place along with top soil by erosion. This water could be stored for providing life saving or protective irrigation to the crops grown in dry land areas. The loss of water takes place in several ways namely run-off, evaporation, uptake through weeds etc. The water could be stored for short period or long period and it can be preserved either in soil, pond or ditches based on situation and utilized for irrigation during dry periods. Amount of rainfall received during each monsoon season by each dry farming districts are discussed below.

Vellore: The average rainfall during North-East monsoon is 353 mm compared to normal rainfall of 362.2 mm. Similarly the South-West monsoon received 442 mm compared to normal rainfall of 435 mm. The length of growing period is 115 days.

Thiruvannamalai: The average rainfall during North-East monsoon is 439.8 mm compared to normal rainfall of 443.4 mm. Similarly the South-West monsoon received 465.8 mm compared to normal rainfall of 451 mm. The length of growing period is 146 days.

Namakkal: The average rainfall during North-East monsoon is 291 mm compared to normal rainfall of 290.7 mm. Similarly the South-West monsoon received 317 mm compared to normal rainfall of 316.2 mm. The length of growing period is 112 days

Erode: The average rainfall during North-East monsoon is 323.5 mm compared to normal rainfall of 321.9 mm. Similarly the South-West monsoon received 213.1 mm compared to normal rainfall of 196.9 mm. The length of growing period is 80 days.

Trichy: The average rainfall during North-East monsoon is 356.1 mm compared to normal rainfall of 385.2 mm. Similarly the South-West monsoon received 270.3 mm compared to normal rainfall of 272 mm. The length of growing period is 94 days

Karur: The average rainfall during North-East monsoon is 365.4 mm compared to normal rainfall of 317.9 mm. Similarly the South-West monsoon received 249.7 mm compared to normal rainfall of 192 mm. The length of growing period is 105 days

Perambalur: The average rainfall during North-East monsoon is 449.6 mm compared to normal rainfall of 475.8 mm. Similarly the South-West monsoon received 349.6 mm compared to normal rainfall of 356.5 mm. The length of growing period is 112 days

Pudukkottai: The average rainfall during North-East monsoon is 418 mm compared to normal rainfall of 399.7 mm. Similarly the South-West monsoon received 350.7 mm compared to normal rainfall of 352.5 mm. The length of growing period is 95 days

Madurai: The average rainfall during North-East monsoon is 373 mm compared to normal rainfall of 406.9 mm. Similarly the South-West monsoon received 305.4 mm compared to normal rainfall of 288.8 mm. The length of growing period is 107 days.

Ramnad: The average rainfall during North-East monsoon is 507.4 mm. compared to normal rainfall of 486 mm. Similarly the South-West monsoon received 136.1 mm compared to normal rainfall of 141.3 mm. The length of growing period is 70 days

Sivagangai: The average rainfall during North-East monsoon is 415.5 mm compared to normal rainfall 390.8 mm. Similarly the South-West monsoon received 289.6 mm compared to normal rainfall of 300 mm. The length of growing period is 93 days

Thoothukudi: The average rainfall during North-East monsoon is 410.1 mm compated to normal rainfall of 433.3 mm. Similarly the South-West monsoon received 86.8 mm compared to normal rainfall of 66.7 mm. The length of growing period is 64 days

Agricultural drought is a condition in which there is no rainfall and insufficient soil moisture availability in soil to the crop.

Under normal condition, excessive moisture is far less problem than drought. Thornth waite defines drought as "a condition in which the amount of water needed for transpiration and direct evaporation exceeds the amount of moisture available in the soil".

1. Permanent Drought - Arid Climate
2. Seasonal Drought - Climate with annual periods of dry weather
3. Drought due to precipitation variability
    a) Moderate drought - lower quality or yield
    b) Severe Drought - Failure of Crop

High Temperature - Growth is slowed or even stopped, regardless of the moisture supply and fruit set affected.

Low Temperature; Most of the crop plants are injured and many killed when night temperature is very low. Tender leaves and flowers are sensitive to low temperature and frost. Rapidly growing and flowering plants are easily killed. Low temperature interferes with the respiration of plants. Low temperature with wet soil, results in the accumulation of harmful products in the plant cells. Frost also interferes with plant metabolism.

Wind

Strong Wind - Mechanical damage to plants lodging, shattering of seeds.

Continuous strong wind interferes pollination by insect, fruits and nuts stripped from trees, depletion of soil moisture and short statured plants completely covered with sand or dust blown by wind.

Aberrations in Rainfall

Aberration means, the deviation from the normal behaviour of the rainfall. As we all know the principal source of water for dry land crops is rain, a major portion of which is received during the monsoon period. Bursts of rain alternated with "breaks" are in common. There are at least four important aberrations in the rainfall behaviour.

- The commencement of rains may be quite early or considerably delayed.
- There may be prolonged breaks during the cropping season (Intermittent drought).
- The rains may terminate considerably early (early cessation of rain) or continue for longer periods.
- There may be spatial and or temporal aberrations.

EARLY OR DELAYED ONSET OF MONSOON

To quantify the aberrations in the onset of monsoon, 50 years of data to be analysed for the dates of onset of monsoon and has to be studied for different regions of the country. For example, it was seen that the normal date of onset of monsoon in the Madhya Pradesh and Maharashtra region is 10th June. In eight percent of the years, onset of monsoon can occur during last week of May (May 28th in 1925) and in 10 percent of the years, it is as much delayed as beyond 21st June. The aberrations require changes in crops and varieties. Crops like Sorghum, Bajra, Pulses and Oilseeds can be grown in Kovilpatti tract of Tamil Nadu. If monsoon is delayed up to late October, bajra, pulses, sunflower can be raised. If it is very much delayed up to first week of November only sunflower can be sown.

BREAKS IN THE MONSOON RAINS (INTERMITTENT DROUGHT)

The breaks can be of different durations. Breaks of shorter duration (5-7 days) may not be a serious concern, but breaks of longer duration of 2-3 weeks or even more, lead to plant -water stress causing reduction in crop production. These breaks, intermittent drought, can be different magnitude and severity and affect different crops in varying degrees. The yield of many drought resistant crops is not seriously affected, but in several sensitive crops, the yield reduction was heavy.

Another aspect of the breaks or intermittent drought is the stage of the crop at which the drought occurs. The effect on crop will be different at different stages.

Another important factor is the effect of breaks or intermittent drought depends on the physical properties of the soil, particularly its water holding capacity. Deep black soils have capacity to store as much as 300 mm of available soil moisture in one meter depth; whereas, light soils like desert soil can store only as little as 100 mm or so. Hence, drought is more pronounced in the soils having less storage capacity.

EARLY WITHDRAWAL OF MONSOON

For example, the normal withdrawal of SWM in Rayalaseema region will be between September 25th and October 15th. But in four percent of the years out of 55 years, monsoon can withdraw during first fortnight of September and in 10 percent of the years, it withdraws during the month of December.

Since, crops and varieties in any given region are selected based on the normal length of the growing season. Persistance of rains, much beyond normal dates creates an extraordinary situation.

Under Kovilpatti (Tamil Nadu) condition, short duration bajra and sunflower will be suitable under early withdrawal of monsoon.

Cultural practices to mitigate the effect of moisture stress due to intermittent drought and early withdrawal of monsoon are 1.shallow interculture to eradicate weeds, 2. Maintain soil mulch to conserve soil moisture, 3. Application of surface mulch, 4. Thinning of crops by removing alternate rows as in sorghum, 5. Recycling of stored run off water, 6. Ratooning in crops like sorghum and bajra, 7. For indeterminate crops like castor and redgram give 2-3 percent Urea spray after a rain.

Uneven distribution of monsoon rains, in space and time over different parts of the country:

Such situation are encountered almost every year in one or another part of the country during monsoon period leading to periodical drought and flood situations.

High variability of rainfall (or more precisely the soil water) is the single factor which influences the high fluctuations in the crop yields in the different parts of the country.

HEAT WAVE

A region is considered to be in the grip of moderate heat wave when it recorded maximum temperature exceeding the normal by 5 to

8^0C. Heat wave is common in UP, 54 percent probability in the month of June. Incidences are maximum in Western UP. Persistance is 5-6 days particularly more in June.

Effect of Heat Wave

Thermal death point affects the photosynthesis and respiration. Increased respiration, depletion of reserve food, Sun clad, stem girdle are some of the effect of heatwave in the plants.

Soil Temperature

In many cases, soil temperature is more important to plant life than air temperature. It influences the germination of seeds and root activities. It influence the soil-borne diseases like seedling blight, root rot etc. The decomposition of organic matter will be higher in higher soil temperature. It controls the nutrient availability. In tropics, high temperature of soil causes regeneration of potato tubers. It also affects nodulation in legumes.

COLD WAVES

A region is said to be in the grip of a moderate cold wave when it recorded minimum temperature falls short of the normal by 6 to 8^0C. Generally experienced from November to March. Severe cold wave generally prevail from January to March in western UP.

Some Other Aberration Terminologies

STORM

A marked atmosphere disturbance characterized by a strong wind, usually accompanied by rain, snow, sleet (rain that freezes as it falls - mixture of rain with snow or hail) or hail and often thunder and lightning.

Thunder Storm

A storm invariably produced by a cumulonimbus strong wind gusts, heavy rain and sometimes hail. It is usually of short duration, seldom over two hours.

- Vertical motion is having many weather modification.
- Upward motion results due to expansion than it gets cooled and eventually condensation.

- Cumulonimbus cloud types are closely related to the strength of the vertical motion.
- A thunderstorm is as the name implies a storm accompanied by thunder and therefore lightning. As Benjamin Franklin demonstrated in 1750, lightning discharges giant electrical sparks.
- Cumulonimbus clouds, therefore are great electrical generators. The cloud produce '+' and '-' value, charges by charged poles. The lower part of the cloud is negatively charged and upper part is positively charged.

Hail

Precipitation in the form of balls or irregular lumps of ice.

HAIL STORM

Small, round pieces of ice (hail) that sometimes fall during thunder storms (frozen rain drops, hail storms).

- Hails may be sometimes greater in size than a large marble.
- It falls from cumulonimbus clouds.
- Hails are destructive to crops -mechanical damage to crops, structures etc.

HURRICANE

A violent tropical cyclone with wind speed of 73 or more miles per hour usually accompanied by torrential rain fall (very heavy), originating usually in West Indian Region.

TORNADO

Tornado - Spanish word - Tornas means "to turn"

- The smallest vortex (Whirlpool, whirl or powerful eddy of air, whirl wind - a whirling mass of water forming a vacuum at its centre, into which anything caught in the motion or drawn).
- Eddy - current of air, water, etc. moving against the main current and with circular motion.
- Most powerful one.
- The intense rotation is confined normally to a diameter of a kilometer or less.
- Its wind speed can reach even 300 kmph.

WATER SPOUTS

- The tornado occasionally forms over water.
- Because of high moisture content of the air, the funnels are heavily loaded with water drops, so, they look like a stream of water pouring from the base of the cloud. For this reason they are called water spouts.

DUST DEVIL

A whirl wind that frequently form on very hot days especially over desert is the dust devil. Normally, there are no clouds associated with it.

Cyclone means closed circulation about a low pressure centre which is anti clockwise in the Northern Hemisphere.

- Cyclonic whirls are the 'storms' of middle latitude.
- In the temperate latitude they produce much of the Winter precipitation.
- Around the low pressure centers.
- Air circulates anti clockwise direction in Northern Hemisphere
- The air is heterogenous in relation to temperature and moisture.

CYCLONE AND ANTICYCLONE

The direction of circular motion of anticyclones is clockwise in Northern hemisphere and anti clockwise in Southern hemisphere.

- This circulation subside whirling @ 10 -15 cm/sec, and fair weather generally prevail.
- The air masses are homogenous with respect to temperature and moisture.

TYPHOON

Any violent tropical cyclone originating in the Western pacific, especially in the South China sea.

Frost

When the dew point is below 0° C, moisture passes directly from the gaseous to solid state. This results in the formation of ice crystals, called frost. Frost occurs mostly in lower places like valleys of the mountain where there is no outlet. The cold heavy air drains along the sloping surface with such low places create temperature inversions.

CHAPTER - 13

El Nino and La Nino

El Nino means 'The Little One' or 'Christ Child' in Spanish. It was originally recognized by fisherman of the coast of South America as the appearance of usually warm water in the Pacific Ocean, occurring near the beginning of the year i.e. around Christmas. That is why the name is used. El Nino, the periodic warming of Pacific Ocean waters that affects the weather worldwide. C.Fred and T.Andrus of the University of Georgia studied the ancient fish bones from refuse left about 6000 years ago by ancient peoples in Peru and found that ocean catfish lived in water that averaged 60 to 70 warmer than now and that there was little variation in temperature. He concluded that if El Nino was occurring at the modern rate, once every two to seven years, then the bones from the fish would have reflected the temperature variation.

El Nino results from interaction between the surface layers of the ocen and the overlying atmosphere in tropical Pacific. It is the internal dynamics of the coupled ocean-atmosphere system that determine the onset and termination of El Nino events. The physical processes are completed, but they involve unstable air-sea interaction and planetary scale oceanic waves. The system oscillates between warm (El Nino) to neutral conditions with a natural periodicity of roughly three to four years. External forcing from volcanic eruption has no relationship with El Nino.

Every time El Nino event has different in magnitude and in duration. Variation in the Southern Oscillation Index (SOI) can give the magnitude of the El Nino. The great width of the Pacific Ocean is the main reason to felt the El Nino Southern Oscillation (ENSO) events in that ocean as compared to the Atlantic and Indian Oceans. Most current theories of ENSO involve planetary scale equatorial waves. The time it takes these waves to cross the Pacific is one of the factors that sets the time scale and amplitude of ENSO climate anomalies. The narrower width of the Atlantic and Indian Oceans means the waves can cross those basins in less time, so that ocean adjusts more quickly to wind variations. Conversely, wind variations in the Pacific Ocean excites waves that take a long time to cross the basin, so that the Pacific adjusts to wind variations more slowly. This slower adjustment time allows the ocean-atmosphere system to drift further from equilibrium than in the narrower Atlantic or Indian Ocean, with the result that international climate anomalies are larger in the Pacific.

There is another way in which the width of the Pacific allows ENSO to develop there as compared to the other basins. In the narrower Atlantic and Indian Oceans, bordering land masses influence seasonal climate more significantly than in the broader Pacific. The Indian Ocean in particular is governed by monsoon variations, under the strong influence of the Asian land mass.

Due to advancement in the technology, a day by day account of events can observed from satellites and from sensors in the ocean itself. Such advancement has not been available earlier. This advancement in the technology provides the high definition information on the Tropical Ocean and atmosphere. It will give lot of interested particulars on El Nino and its effect on the livings on the earth.

La Nino: It means 'The Little Girl. It is some times called as El Viejo, anti-El Nino or a cold event or a cold episode. La Nino is characterized by unusually cold ocean temperatures in the equatorial Pacific. Global climate anomalies associated with La Nino tend to be opposite to El Nino. At higher latitude, El Nino is only one of a number of factors that influence climate. However, the impacts of El Nino and La Nino at these latitudes are most clearly seen in winter period.

Chapter - 14

Phenology and Bioclimatic Law

Plant growth is resultant of all the environmental factors - climatic, physiographic, edaphic and biotic factors.

For a particular field - it is primarily a function of climate with temperature and light - being the most important factors. - Close relationship exists between plant phenology and both latitude and altitude.

DEGREE DAYS

At a given location, the period between planting and harvesting is not a specific number of calendar days but rather a summation of energy unit, which may be represented as degree days.

A degree day for a given crop is defined as a day on which the mean daily temperature is one degree above the zero temperature (that is the zero temperature) Spring wheat 32 - 40^{0}F (depending on variety) Oat : 43^{0}F; corn: 54 - 57 ^{0}F, Sweet corn : 50 ^{0}F, Potatoes : 45^{0}F, Peas:40^{0}F, Cotton 62-64^{0}F.

The period required for achieving maturity is also a function of the length of day (photoperiod). Crops planted in early in the spring require more calendar days to mature than the same crops planted later.

Phenology

The study of natural phenomena that recur periodically (as migration, blossoming etc.) and their relation to climate and changes in season is called Phenology.

USES OF STUDIES OF PHENOLOGY

- Studying the phenology of crops over a number of years make it possible to establish planting schedules that are not adapted to local climate, availability of labour and market demands.
- It is also possible to avoid periods of maximum hazard from insects, diseases or seasonal weather phenomena.
- In irrigated areas phenological data can be used more and it is the basis for determining the best time to plant in order to make the least demand for water.

Ex. Nendran Banana - Jan-Feb

IR 50 Rice - Only in Kuruvai - not in Samba /Thaladi,

Brown plant hopper in rice - Oct - Nov. more population -Do not grow susceptible varieties.

Red hairy caterpillar - Emergence of insect from pupa during rainfall occurrence particularly in Summer.

High temperature - pest population will be more.

High RH & low temp - Powdery mildew in bajra.

BIOCLIMATOLOGY

The science deals with the effects of climate on living matter.

Phenology

It is the science which relates climate to periodic events in plant and animal life.

Phenological date for crops include facts such as dates of planting, germination and emergence of ripening and dates of harvest.

- These depend on climatic conditions preceding each event as well as climatic factors at the time of occurrence of the event. Their specific relationship to climatic elements is not fully understood.
- Generation of farmers have kept phenological diaries listing the dates of observed development of natural vegetation and crops as well as facts concerning the periodic reactions of birds and animals to climate and the seasons.
- Such information - chief basis for denoting "Signs of spring" and "Signs of autumn" and so on. Phenological maps can be

prepared with isospheres connecting places at which a phenological event takes place on the same date. The difference in dates of significant phases of plant development over a given distance in the phenological gradient.

Heavy Rains

- Directly damage plants.
- Interfere with flowering and pollination.
- Surface soil making seedling emergence difficult.
- Lodging, difficult the harvest.
- Grain susceptible to spoilage and disease.

Snow, sleet, freezing rains, threats to winter plants - breaking of branches in trees and shrubs. Suffocation of crop plants such as winter wheat.

Hail-may be disaster - direct damage to plants. It is dependant on stage of crop and intensity of hail storm. It damages the young plants, shedding of leaves, flowers and ear heads.

Chapter - 15

Agricultural Seasons of India

Season is a period in a year, comprising few months during which the prevailing climate does not vary much. Growing season for a crop is more important for its yield and other management practices, to be followed.

Indian Meteorological Department has divided the year into four seasons.

i.	Summer	-	March-May
ii.	Monsoon	-	June-September
iii.	Post Monsoon	-	October -November
iv.	Winter	-	December-February

The monsoon season is designated as ***kharif,*** whereas the post monsoon and winter seasons are together designated as Rabi throughout the India.

Based on temperature ranges, three distinct crop seasons have been identified in India.

i. Hot weather (Mid February - Mid June)

ii. *Kharif* or Rainy season (mid June to Mid October)

iii. Rabi (Mid October to Mid February)

In Southern stats (Tamil Nadu, Andhra Pradesh and Karnataka) there is slight variation in the seasons based on the rainfall duration as

i. Winter - January and February

ii. Summer - March to May

iii. Rainy Season - a) South West monsoon – June to September

b) North East monsoon - October to December

Based on the criteria, monthly precipitation and temperature, the growing season is broadly divided as follows.

i. Hot month - If the average temperature is above 20^0C.

ii. Cold month - If the mean temperature is between 0-15^0 C.

iii. Warm month - If the mean temperature is between 15-20^0 C.

AGRONOMIC CONCEPTS OF THE GROWING SEASONS

Agronomically, the growing season can be defined as the period when the soil water, resulting mainly from rainfall, is freely available to the crop. This condition occurs when the water consumed by the crop is in equilibrium with rainfall and water storage in the soil.

The growing season for a rainfed crop involves three distinct periods during which the soil moisture conditions depend on the rainfall received.

a. **Pre-humid period:** During this period the precipitation will always remain lower than the potential evaporation for the corresponding period. This period corresponds to the sowing period of the crop. Sowing can be done when the precipitation during the week is > 0.5 PET.

b. **Humid period:** During this second period, the precipitation remains higher than the PET. The crops in this period will be in active vegetative and flowering phase and the water requirement will be at its peak. At the end of this period water balanced is on the positive side and the water storage in the soil is on the increase, since the rainfall is higher than the water needs.

c. **Post-humid period:** This period follows the humid period. During this period there is a gradual reduction in the water stored in the soil due to the utilization by the crop plants. The crop will also make use of the rainfall received. This period usually coincides with maturity stage of the crop.

TYPES OF GROWING PERIOD

There are four types of growing period.

1. Normal

In this type, rainfall is in excess during the humid period. At the end of the pre-humid period when precipitation is higher than 0.5 PET sowing of crops are taken up. This type of growing season is prevalent in semi-tropics.

2. Intermediate Type

The precipitation is lower than PET all round the year. The growing season is limited to the period when rainfall is in excess of 0.5 PET. Only drought hardy crops like pearl millet, castor etc., can be grown. Dry farming is highly risky.

3. All Year Round Humid

In this type, the precipitation is more than PET all round the year, indicating the moisture sufficiency for cropping. This type occurs in high rainfall area and mostly perennial crops are raised.

4. All Year Round Dry

The precipitation is lower than 0.5 PET throughout the year. Cropping is not possible in these areas. This type of growing season is found in extremely arid area, mostly the deserts.

The fluctuations in the crop yield depend on the following conditions.

- The length of the rainy season i.e. from sowing to the end of the rains is shorter.
- The quantity and distribution of rain during the pre-humid and humid periods.
- The excess rainfall during humid period should go to soil storage. It may cause water logging and crop lodging.
- The amount of rainfall received during post humid season, may supplement the soil moisture during maturity. This may favourably influence the yield. In India four cropping seasons have been identified by IMD in dry farming area.

S.No.	Name of the Season	Duration (weeks)	Water need met from RF	Crops
1.	Short duration	Upto 10	75%	Very short duration crops.
2.	Medium duration	10-15	75%	Medium duration crops with intercrops.
3.	Extended medium duration	15-20	75%	A medium duration crop followed by short duration crops if soil type is suitable.
4.	Long duration	20-30	75%	Medium duration crops followed by short duration crops.

Common Questions and Answers

Why do we feel hungry very frequently during winter?

When the ambient temperature drops below certain value, the body generates heat by increasing its basal metabolic rate in order to keep up the body temperature. There are two ways by which this metabolic heat is achieved - one, by increasing the breakdown of body's stored fat, and two, by providing the body more fuel to burn in the form of food. Alternately, we can consume more food to meet the increased metabolic demands of the body. The main purpose of intake of food is to supply energy for physical needs. The body maintains a balance between energy expenditure and calorie intake which helps in proper maintenance of body weight.

The food we eat is utilized by the body by a process called metabolism (the various chemical reactions occurring in the body's cells to break down food to give us energy and heat), and this works best at normal body temperature (37^0C). A part of our brain, called hypothalamus, controls this automatic regulation. The hypothalamus functions as a thermostat, and it has two discrete centers for regulating the food intake – a 'feeding centre' and a 'satiety centre', both of them together maintain a judicious balance of feeding behavior. The chief factors which influence these centres are – the body weight, the amount of food present in the gut, the amount of glucose in the blood and finally the body temperature. Thus in cold weather the feeding centre is stimulated so that food intake is increased. It is worth noting here that warm weather decreases the appetite to some extent. However, ambient temperature is not a major player in regulating the food intake in human beings, whereas body weight plays a significant role.

It is interesting to note that we lose appetite in fever, because raised temperature suppresses feeding center.

CHAPTER - 16

Agroclimatic Zones

Climate, in general, is the totality of weather during a longer period and over wider area. Agro climate can be defined as the conditions and effects of varying weather parameters like solar radiation, rainfall etc., on crop growth and production.

Climatic classification is a method of arranging various data of climatic parameters to demarcate a country or region into homogenous zones i.e., places having similar conditions.

ADVANTAGES OF AGROCLIMATIC CLASSIFICATION

- This would enable in exploring agricultural potentiality of the area.
- Locating similar type of climatic zone will enable in identifying the specific problems of soil and climate related to agriculture.
- This will help in introduction of new crops from other similar areas. E.g. Introduction of oil palm in Kerala from Malayasia.
- Development of crop production technologies specific to the region.
- To take up research work to solve the regional problems, and
- To transfer the technology easily among the farmers.

AGROCLIMATIC ZONES OF INDIA

Krishnan and Mukhtar Singh (1968) demarcated soil climatic zones of India by superimposing the moisture index and mean air temperature isopleths on a soil map of India showing major soil types.

Mean annual PE values were computed by Thornthwaite's method and P is mean annual precipitation. The scale adopted in defining climatic zones in terms of moisture indices is:

$$\text{Moisture index} = \frac{P-PE}{PE}\times 100$$

Zone	Moisture index	Moisture belt
1	<-80	Extremely dry
2	-60 to -80	Semidry
3	-40 to 60	Dry
4	-20 to -40	Slightly dry
5	0 to -20	Slightly moist
6	0 to + 50	Moist
7	+50 + 100	Wet
8	> +100	Extremely wet

The classes in terms of temperature are:

Zone	Mean annual temperature (^{0}C)	Temperature belt
A	>28	Very hot
B	25 to 28	Hot
C	20 to 25	Mild
D	10 to 20	Cold
E	<10	Very Cold

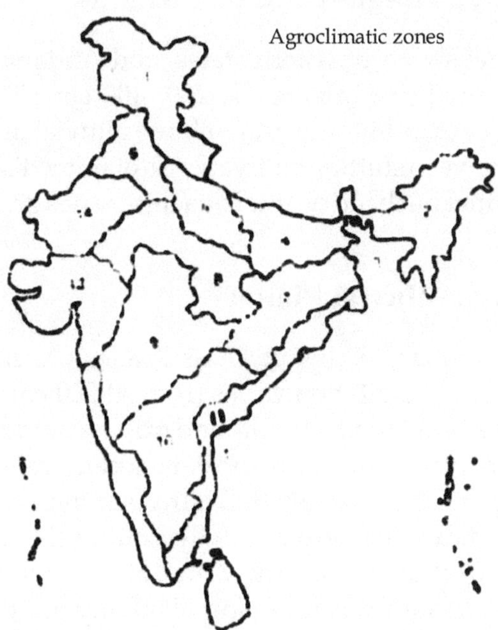

1. Western Himalayas
2. Eastern Himalayas
3. Lower Gangetic plains
4. Middle Gangetic plains
5. Upper Gangetic plains
6. Trans Gangetic plains
7. Eastern Plateau and Hills
8. Central Plateau and Hills
9. Western Plateau and Hills
10. Southern Plateau and Hills
11. East coast Plains and Hills
12. West coast Plains and Hills
13. Gujarat Plains and Hills
14. Western dry regions
15. Islands zone

1. Humid Western Himalayan Region

This region consists of Jammu and Kashmir, Himachal Pradesh and the hill divisions of Kumaon and Garhwal of Uttar Pradesh. The climate varies from hot and subhumid tropical in the southern low tracts to temperate cold alpine and cold arid in the northern high mountains. The annual precipitation ranges from 8 cm in Ladakh to 350 cm in the hills of Himachal Pradesh and Uttar Pradesh. The major groups of soils, namely mountain meadow soils, sub-mountain meadow soils and brown hills soils occur in this region. The region has great potentialities for further development of horticulture and orchards. Deforestation, overgrazing, soil erosion and acidity are the major management issues of the region.

2. Humid Bengal-Assam Basin

This region represents the Ganga-Bramhaputra alluvial plain and deltaic deposits. The region has a hot humid monsoon climate. The annual precipitation ranges from 220-400 cm. Management of catchments areas to prevent the occurrence of frequent flood is the major management issue. The predominant soil groups are alluvial, red, brown, hill and coastal soils.

3. Humid Eastern Himalayan Region and Bay Islands

This region comprises the six northeastern states, and Andaman and Nicobar Islands. The annual precipitation is 200 -400 cm. The predominant soil groups are brown hill, red and yellow, alluvial and laterites. Management alternative to shifting cultivation is of immediate concern. Development of animal husbandry and aquaculture has great potentialities.

4. Subhumid Sutlez –Ganga Alluvial Plains

This region consists of Punjab, plains of Uttar Pradesh, Delhi territory and Bihar. The annual precipitation varies from 30-200 cm in the plains. Hot summer and cold winter prevail. Ground frost is common in December and January. The numerous hills torrents, locally called 'chos' cause severe damage to the land in the rainy season. The predominant soil groups are calcareous sierozem in the southwest, red soil in the sub mountain zone, alluvial soils (new and old alluvium) and patches of saline and alkali soils. Salinity and alkalinity are the main soil problems in Punjab and Uttar Pradesh. North Bihar is badly affected by floods. In the plateau, soil erosion due to deforestation and overgrazing is serious.

5. Subhumid to Humid Eastern and Southeastern Uplands

This region consists of Odisa, Andhra Pradesh and eastern Madhya Pradesh. It is characterized by undulating topography, denuded hills, plateau, river valleys, high lands of the Eastern Ghats and wide basins of Chattisgarh. The annual rainfall varies from 100-180 cm. The predominant soil groups are mixed red and black, red and yellow, red, sandy, laterite, black and alluvial (river and coastal sides). Soil salinity and flood in the coastal deltas are major management problems. The red and laterite soils are mostly acidic and require liming.

6. Arid Western Plains

This region covers the states of Haryana, Rajasthan and Gujarat and the Union Territory of Dadra and Nagar Haveli. It is an alluvial plain with sand dunes, sandy plain, and granite hills intersected by the Aravalli range. The annual precipitation varies from 10 cm in the extreme west to 65 cm in the extreme east. The predominant soil groups are alluvial, black, desert, saline and alkali soils. The frequent dry spells and combating desertification are major management issues.

7. Semiarid Lava Plateau and Central Highlands

This region consists of Maharastra, Western and central Madhya Pradesh and Goa, Daman and Diu. The region has a predominant plateau, undulating land surface in Madhya Pradesh, part of Western Ghat and Coastal alluvial plains. The annual precipitation varies from 70-125 cm in the plateau and plains. The major soil groups are black, alluvial, mixed red and black, yellowish brown and laterites. The crops grown in light textured soils in the central part suffer from drought. Soil erosion is a serious problem.

8. Humid to Semiarid Western Ghats and Karnataka Plateau

The region consists of Karnataka, Tamil Nadu, Kerala Pondicherry and Lakshadweep Islands, and can be divided into western Ghats, plateau, river valleys and coastal plains. The rainfall varies from 60 cm in the plains to 300 cm in the Western Ghats and west coast. The important soil groups are black, red, laterite and alluvial. The management issues of major concern are drought, soil erosion, acidity, salinity and alkalinity. Development of horticulture including arid horticulture and plantation crops has great scope.

The International Crop Research Institute for Semi Arid Tropics (ICRISAT) adopted the Troll's (1965) classification for classifying climate of India. The Semi Arid Tropics (SAT) of India shows large areas of Rajasthan, Gujarat and parts of Madhya Pradesh in dry SAT (2 to 4.5 humid months). This region consists of 26 per cent of the total area of tropical India and represents 47 per cent of India SAT. The wet-dry zone having 4.5 to 7 humid months, 200 to 500 km wide belt running from northwest to south India around the dry semiarid zone. It contributes to 30 per cent of the total area of tropical India and 53 per cent of the SAT of the country. The total SAT area amounts to 56 per cent of total tropical India. This figure is very close to the estimate of 54 per cent made by the Economic Department of ICRISAT. If Troll's classification is accepted, most area of the Indian SAT falls between 600 and 1400 mm isohyets of annual rainfall.

In 1976, National Commission on Agriculture carried out the climatic delineation of India by examining the monthly rainfall distribution at all provincial raingauge stations of India. They chose limits that have a closer relation to the broad requirements of crops. Since the time span of most of the crops is usually 90 days or more, the limites set by them were:

- Rainfall greater than 30 cm per month for at least 3 consecutive months would be suitable for a crop like paddy which needs more water.
- Twenty to thirty cm per month for less than 3 consecutive months would be suitable for crops that needs high amount of water but less than that of paddy (maize and black gram),
- Ten to twenty cm per month for atleast 3 consecutive months is considered suitable for crops requiring less water (Pearl millet and small millets),
- Five to ten cm per month is just sufficient for crops that have low water requirement (field beans and ephemeral grasses), and
- Rainfall less than 5 cm per month is not of much significance for crop production.

In India, rain fall occurs variably in quantity and distribution. Based on the quantity received, symbol was given. Accordingly, the adopted symbolic representation for rainfall levels was

Symbol	Monthly rainfall (cm)
A	>30 cm
B	20-30
C	10-20
D	5-10
E	<5 cm

February –May / (June – September) / October – January

Pre monsoon months / SouthWest monsoon months / Post monsoon months

In denoting the year's distribution of rainfall, the southwest monsoon months from June to September are given the central position in brackets. To the right is the distribution for the postmonsoon months from October to January and to the left is that of premonsoon months, Feruary to May. The number of months in the respective season in which monthly rainfall is within the specific limit mentioned above is indicated by a number as a subscript to the symbols given above.

Maps showing different rainfall categories and interrelated rainfall categories were drawn for various states and combined with tahsil ('country') cropping patterns and relative yield indices of crops of different states by the National Commission on Agriculture.

Planning Commission of India (1989) made an attempt to delineate the country into different agro climatic regions based on homogeneity in rainfall, temperature, topography, cropping and farming systems and water resources. India is divided into 15 agro climatic regions

During the year 1989, the Planning Commission made an attempt to delineate India into different agroclimatic zones. Based on the similarity in rainfall, temperature, soil topography, cropping, farming system and water resource, India has been divided into different agro-climatic regions. This was done mainly to identify the production constraints and to plan future strategies.

The country has been divided into 15 agroclimatic zones. This classification is suggested for purpose of cropping for balanced regional growth.

Region	Areas
1	Western Himalayas
2	Eastern Himalayas
3	Lower Gangetic plains
4	Middle Gangetic plains
5	Upper Gangetic plains
6	Trans Gangetic plains
7	Eastern Plateau and Hills
8	Central Plateau and Hills
9	Western Plateau and Hills
10	Southern Plateau and Hills
11	East coast Plains and Hills
12	West coast Plains and Hills
13	Gujarat Plains and Hills
14	Western dry regions
15	Islands

1. Western Himalayan Zone

This zone consists of three distinct sub-zones of Jammu and Kashmir, Himachal Pradesh and Uttar Pradesh hills. The region consists of hill soils of cold region, podsolic soils, mountain medium soils, hilly brown soils, lands of the region have steep slopes in undulating terrain. Soils are generally silty loams and these are prone to erosion hazards.

2. Eastern Himalayan Zone

Sikkim and Darjeeling hills, Arunachal Pradesh, Meghalaya, Nagaland, Manipur, Tirupura, Assam, Mizoram and Cooch district of

West Bengal, falls under this region of high rainfall and high forest cover, shifting cultivation practiced in nearly one-third of the cultivated area and this has caused degradation of soils with the heavy run off, massive soils erosion and floods in the lower reach and basin.

3. Lower Gangetic Plains Zone

This zone consists of West Bengal - lower Gangetic plains regions. The soils are mostly alluvial and are prone for floods.

4. Middle Gangetic Plains Zone

This zone consists of 12 districts of eastern UP and 27 districts of Bihar plains. This zone has a geographical area of 16 million hectares and rainfall is high. About 39 per cent of gross cropped area are irrigated and the cropping intensity is 142 per cent.

5. Upper Gangetic Plains Zone

This zones consists of 32 districts of UP, irrigation is through canals and tube-wells. A good potential for exploitation of ground water exists.

6. Trans Gangetic Plains Zone

This zone consists of Punjab, Haryana, and Union territories of Andaman and Nicobar and Delhi and Chandigarh and Sriganganagar district of Rajasthan. The characteristics of this area are higher net sown area. Higher irrigated area high cropping intensity and high ground water utilization has witnessed remarkable agricultural revolution; still there is possibility to increase the productivity.

7. Eastern Plateau and Hills Zone

This zone consists of Eastern part of Madhya Pradesh. Southern part of West Bengal and most of Island Odissa, the soil are shallow and medium in depth and the topography is undulating with a slope of 1 to 10 per cent. Irrigation is through tanks and tube wells.

8. Central Plateau and Hills Zone

This zone consists of 46 districts of Madhya Pradesh and Uttar Pradesh and Rajasthan.

9. Western Plateau and Hills Zone

This zone comprises the major part of Maharashtra, parts of M.P. and southern districts of Rajasthan. The average rainfall of the zone is

904 mm. The net sown area is 65 per cent and forests occupy 11 per cent. The irrigated area is only 12.4 per cent with canals as the main source.

10. Southern Plateau and Hills Zone

This zone comprises of 35 districts of Andhra Pradesh, Karnataka and Tamil Nadu, which are typically semi arid zones. Dry land farming is adopted in 81per cent of the area and the cropping intensity is 100 per cent.

11. East-Coast Plains and Hills Zone

This zone comprises of East coast of Tamil Nadu, Andhra Pradesh and Odissa. Soils are mainly alluvial and coastal sands. Irrigation is through canals and tanks.

12. West Coast Plains and Hills Zone

These zones comprised the West coast of Kerala, Karnataka, Maharashtra and grows with variety of crop patterns, rainfall and soil types.

13. Gujarat Plains and Hills Zone

This zone consists of 19 districts of Gujarat. This zone is with low rainfall in most part and only 32.5 per cent of the area are irrigated, largely through well and tube wells.

14. Western Dry Regions Zone

This zone comprises of 11 districts of Rajasthan and characteristic by hot sandy desert erratic rainfall, high evaporation and scanty vegetation. The ground water is deep and often brackish famine and drought is the common features of the region.

15. Island Zone

This zone covers the islands territory of Andaman and Nicobar and Lakshadeep which are typically equinox with rainfall of 3000 mm spread over eight to nine months. It is largely a forest zone with undulated levels.

AGROCLIMATIC ZONES OF TAMIL NADU

Geographical location of Tamil Nadu $8^{0}5'$ and $13^{0}10'$ North latitudes $76^{0}15'$ and $80^{0}20'$ East longitudes, Coastal line, approximately, 1000 km climate: Temperature: range: Maximum 29 -38^{0}C and Minimum 19-27^{0}C.

Efficient Crop Zones

The new efficient crop zone approach should aim in utilizing the natural resources to the fullest extent. The uneconomical crops should be replaced by more economical one. The crops should create more employment opportunities and economic stability of the farmers.

Based on the productivity, efficiency of the crops, each state has been divided into five categories.

1.	Efficient zone	:	The productivity of the crops is high and also stable due to the prevalence of the optimum conditions.
2.	Potentially efficient zone	:	The productivity is high but unstable.
3.	Moderately efficient zone	:	Stable, medium productivity.
4.	Less efficient zone	:	Unstable, medium productivity.
5.	Inefficient zone	:	Low productivity.

Based on soil characteristics, rainfall distribution, irrigation pattern, cropping pattern and other ecological and social characteristics, the State Tamil Nadu has been classified into seven agro-climatic zones. The following are the seven agro-climatic zones of the State of Tamil Nadu.

1. Cauvery Delta zone
2. North Eastern zone
3. Western zone
4. North Western zone
5. High Altitude zone
6. Southern zone and
7. High Rainfall zone

1. Cauvery Delta Zone

This zone includes Thanjavur district, Musiri, Tiruchirapalli, Lalgudi, Thuraiyur and Kulithalai taluks of Tiruchirapalli district, Aranthangi taluk of Pudukottai district and Chidambaram and Kattumannarkoil taluks of Cuddalore and Villupuram district. Total area of the zone is 24,943 sq.km. in which 60.2 per cent of the area i.e., 15,00,680hectares are under cultivation. And 50.1 per cent of total area of cultivation i.e., 7,51,302 hectares is the irrigated area. This zone receives an annual normal rainfall of 956.3 mm.It covers the rivers of Cauvery, Vennaru, Kudamuruti, Paminiar, Arasalar and Kollidam.The major dams utilized by this zone are Mettur and Bhavanisagar. Canal irrigation, well irrigation and lake irrigation are under practice. The major crops are paddy, sugarcane, cotton, groundnut, sunflower, banana and ginger. Thanjavur district, which is known as "Rice Bowl" of Tamilnadu, comes under this zone.

2. North Eastern Zone

The area covered by this zone is Chengleput district, North Arcot district, Cuddalore and Villupuram district excluding, Chidambaram and Kattumannarkoil taluks and Ariyalur. The total geographical area of the zone is 31,194 Sq. kms. The area under cultivation is 50.5 per cent of the total area of cultivation. The annual normal rainfall is 1109 mm. The major rivers are Polar, Ponniar, Cheiyear, Vellar, Thenpennai, Manimuthar and Komugi. Major dams used for irrigation are Mettur, Sathur, Veedur, Komugi, Manimuthar and Wellington. Canal irrigation, well irrigation, Irrigation by lakes and by dams is under practice. The major crops are paddy, cholam, cumbu, ragi, groundnut, sugarcane and cashewnut. Chengleput district which is known as "Lake District" and also where the popular "Madhurandhagam lake" comes under this zone.

3. Western Zone

This zone covers Periyar district, Coimbatore district, Tiruchengode taluk of Salem district and Northern part of Madurai district. This zone constitutes an area of 15,678 sq.km. The area under cultivation is 6,98,105 hectares which is 44.5 per cent of the total area. And only 42.4 per cent of the area under cultivation is the irrigated area i.e., 2,96,201 hectares. The annual normal rainfall is 653.7 mm. Cauvery, Noiyal, Bhavani, Uppar, Sirvani and Amaravathi are the major rivers and Mettur, Bhavanisagar and Amaravathi are the major dams utilised by the zone.

4. North Western Zone

This zone covers Dharmapuri district excluding hilly areas, Salem district excluding Thiruchengodu taluk and Peramabalur district. This zone covers an area of 18,271 Sq. kms in which 10,28,097 hectares, which is 56.3 per cent, is under cultivation. Out of total area of cultivation, only 23 per cent i.e., 2,35,828 hectares is the irrigated area. The annual normal rainfall of the zone is 849 mm. This zone has been identified as moderately drought prone. The major rivers of this zone are Cauvery, Thenpennai and Manimuthar. Mettur and Krishnagiri are the major dams in this zone. Paddy, wheat, maize, ragi, bajra, sugarcane, groundnut, cotton, sunflower, tobacco and mango are the major crops of this zone. Forest area in this zone constitutes nearly 30 per cent i.e. 5,35,282 hectares of the area of the zone, which is nearly 25 per cent of the total forest area of the State.

5. High Altitude Zone

This zone covers the Nilgiris, Kodaikanal, Shevroy, Elagir Javadhi, Kollimalai, Pachamalai, Yercaud, Anamalais, Palani and Podhigaimalai. This zone covers an area of 2,549 sq.kms. The area under cultivation is 73,689 hectares, which is only 28.9 per cent of the total geographical area of the zone. Furthermore, only 0.84 per cent of the total cultivated area is the irrigated area i.e., 621 hectares. The annual normal rainfall is 1857 mm. There are no dams for irrigation in this zone since there are no major rivers. Paddy and groundnut are cultivated relatively in less extent. The major crops are tea, coffee and vegetables. Forest area is 1,50,139 hectares which is 58.9 per cent of the total geographical area of the zone.

6. Southern Zone

This zone comprises the districts of old Ramanathapuram, Nellai Kattaboman, V.O. Chidambaranar, Kamarajar and Dindigul district, Natham, Melur, Thirumangalam, Madurai South and Madurai North taluks of Madurai districts and Puddukottai district excluding Aranthangi taluk. This zone constitutes an area of 36,655 sq.kms. The total area under cultivation is 16,50,250 hectares which is 45 per cent of the total area.

Around 44 per cent of irrigated area under cultivation is 7,22,166 hectares. It covers the rivers of Vaigai, Sitrar, Thamraparani, Numbiar, Pachaiyar, Kludar, Arjunar, Kodumudiyaar, Manimuthar, Periyar and Vaigai. The dams used by this zone are Periyar, Vaigai, Manjalar and Bhabanasam. The patterns of irrigation are well irrigation, canal irrigation, irrigation by dams and by lakes. Paddy, cholam, cumbu, ragi, groundnut, cotton, banana and tobacco are the major crops. This zone is prone to frequent drought. The annual normal rainfall is 816.5mm.

7. High Rainfall Zone

This zone consists of only Kanyakumari district. The total geographical area is 1,684 Sq. kms, in which 63.1 per cent of the area i.e. 1,06,260 hectares are the area under cultivation. And 47.4 per cent of the total area under cultivation is the irrigated area i.e., 50,349 hectares. This zone receives an annual normal rainfall of 1456mm. It covers the rivers of Pazhaiyar, Kothair and Paraliyar. And the dams utilised by this zone are Pechiparai, Periyar, Sitharu Dam-1 and Sitharu Dam-2. The major pattern of irrigation is river irrigation. Paddy, coconut, vegetables, tea, cashew nut, banana and rubber are the major crops.

CHAPTER - 17

Weather Forecasting

Environment in which crops are grown dictates their final yield. Of these environmental factors, climate and weather the uncontrolable factors, have maximum influence on crop productivity. Vagaries of weather subject a crop to different ecological situations from year to year leading to differential responses of crops to input resources. This situation limits the use of costly inputs for realizing optimum yield. While weather itself cannot modify, the hazardous weather can often be mitigated by agronomic management. Therefore, the primary requirement for initiating agronomic measures against weather hazards is foreknowledge of weather situation that is likely to develop in an area.

WEATHER FORECAST REQUIREMENTS

For efficient use of inputs in crop production, package of practices have to be decided very early in the season. Once initiated, these cannot be altered. For such an advance planning one needs to know the likely weather on long range basis that is about three months ahead. To a large extent, crop production depends, in our country, on rainfall vagaries. Long range forecasts needed during kharif and rabi are:

Kharif Season

- Onset and withdrawal of monsoon,
- Prolonged rainless period, and
- Occurrence of heavy rains.

Rabi Season

- Rainfall and cold waves during winter,
- Onset of heat waves and strong winds in spring and
- Hailstorms at commencement of summer.

Accurate weather forecasting can help the farmers in realizing optimal yield, by reducing the crop losses. Forecasting aids in-

- Planning for necessary inputs during the season
- Timely land preparation to take advantage of earliest rain for timely sowing
- Selection of crops and cultivars,
- Proper time and method of fertilizer application for its efficient use,
- Predicting pest and disease incidence and frost occurrence for timely action,
- Timing for weed, pest and disease control.
- Planning for protecting the crop from weather hazards, and

Adjustments in crop harvest timing to reduce the losses at harvest. Weather is the major factor which influences agricultural operations and farm production. A substantial portion of a crop is lost due to aberrant weather. The pre-harvest loss may range between 10 and 100 per cent in various crops. In Punjab, a single rainy spell in April -May of 1982 (maturity stage of the crop) damaged the wheat crop worth Rs.1238 million. The post-harvest loss mainly due to rains and excessive humidity was worth Rs.9.42 million.

USEFULNESS OF WEATHER FORECASTING

1. Though the losses due to weather factors cannot be avoided completely, the loss could be minimized by making adjustment with coming weather through timely and accurate weather forecasting. In Punjab, it is estimated that crops worth Rs.176 million could be saved through better weather forecasting.
2. The weather forecasts also provide guidelines for long range seasonal planning and selection of crops suited to the anticipated climatic condition.

3. By forecasting the anticipated rains,
   i) Irrigation from wells can be avoided by which we can save electricity and diesel
   ii) The harvesting could be advanced if the crop is in maturity stage,
   iii) Threshing of harvested produce could be done before rains by which crop losses can be avoided,
4. Loss of seed, diesel, labour and time can be avoided by not sowing in unsuitable weather.
5. Fertilizer losses can be avoided by not applying during unsuitable weather condition for fertilizer application.
6. Similarly pesticide wastage can also be minimized.

TYPES OF WEATHER FORECASTING

There are three types of weather forecasting for agriculture.

1. Short Range Forecast

It is valid for 24-48 hrs. with 70-80 per cent accuracy. Short range forecast emphasizes on temperature, wind velocity and direction, duration of sunshine, time, amount of precipitation and R.H.

Uses (Applicability)

- Scheduling of irrigation
- Adjusting time of agricultural operations
- Protection of plants from frost.

2. Extended Forecast

It is valid within 5 days with 60-70 per cent accuracy. It emphasis on type of weather, sequence of rainy days, normal weather hazards in farming such as strong winds, extended dry or wet spells.

Uses:

- Useful to determine sowing time,
- Useful to determine depth of sowing,
- Planning of irrigation
- Decision on harvesting

- Decision on time of spraying to get higher efficiency
- Management of labour and equipment

3. Long Range Forecast

It is valid up to four weeks.

Forecasts mainly on abnormal temperature and precipitation.

Uses:

- To decide on soil moisture management
- To decide on irrigation scheduling
- Decision on selection of crops
- Decision to manage irrigation with limited water supply
- Decide on cropping pattern and
- To determine crop yield

WEATHER FORECASTING ORGANIZATION

Suitable organizations have been set up in most parts of the world for weather forecasting. Accepted international norms for measuring weather elements and representing them in international code are being adopted by all the participating countries. There are about 300 meteorological observatories of different types, distributed all over India, for the purpose of forecasting.

Synoptic charts and crop weather calendar are the tools for making weather forecasts.

CHAPTER - 18

Crop Weather Modelling

Variation in field crops between years is associated with many factors. This is mainly due to weather, soil and management factors. There is complex reaction of weather variables among themselves as well as with other factors. Therefore, many attempts were made to study the effect of weather variable on crop yield through simulation modeling.

WHAT IS MODEL?

Model is mathematical representation of a system. The process of developing such representation is termed as modelling.

A model is a schematic representation of our conception of a system. In a general term, a model brings into mind the thoughts about the form and functional form of a real object, like children toy, tailors dummies and mack-ups of building and structure to be later constructed in the real forms. Models also construct the objects or situations not yet in existence in real form. A model can also be referred as a representation of relationship under consideration and may be defined as an act of mimicry.

TYPES OF MODELS

1. Simple Statistical or Empirical Statistical Models

This model rely mainly on the statistical techniques such as correlation or regression of the appropriate plant and environmental variables. The regression co-efficients are not necessarily related to the important processes, but estimate the yield alone. Therefore, many studies are required to produce the regression equations necessary for the wide spread application of this kind of models. A great advantage of these simple crop weather models is that they use readily available weather data.

2. Dynamic Crop Simulation Models

Predict changes in crop stages with time. For example, model which predicts soil water content at a certain depth throughout the season or the one which predicts changing number of bolls on a cotton plant with the march of season are dynamic simulation models.

Crop simulation model predicts the final yield and also provides a quantitative information on intermediate steps like daily weight of different plants parts, which is verified through experimentation. The model acts, like a real crop by gradually growing leaves, stems, roots etc., during a season.

In other words, simulation is the process of using a model dynamically by following a system over a time period. A dynamic crop simulation model is most successfully developed by a multi-disciplinary team consisting of Agro meteorologists, Agricultural Engineers, Plant Physiologists, Agronomists, Soil Scientists, Entomologists and Plant Pathologists.

3. Model Based on Physiological and Physical Aspects

There are mechanistic models where plant and soil processes are described with respect to physiological or physical or chemical aspects. For example, N may be taken up from the soil by the root system depending on soil N content and rate of availability of solution to the roots. Thus physical placement of N in the visibility of root system and transformation are important.

4. Phenological Model

That predicts the crop development from one growth stage to another. These prediction are generally based on the accumulated heat limits.

USES OF CROP WEATHER MODELLING

- The models can be used as a research tool in planning alternative strategies for cropping, land and water management practices for a range of agroclimatic conditions.

- Helpful for economists to work out cost benefit ratio analysis.
- It enables plant breeder to develop crop varieties tailored to different agroclimatic conditions.
- Help to make appropriate management decision for production.
- Help to identify most potential area for research.

ADVANTAGES OF CROP WEATHER MODELLING

- Modelling relates plant growth and development from seedling to maturity.
- The variability of growth and development is understood by basic concepts explained on mathematical basis.
- The response of plants to their macro and micro - environments are quantified.
- It provides an understanding of the development process in plants and also helps in knowing missing data to have complete picture of the processes.
- It will give new ideas leading to experimental approaches.
- Modelling enables the researches to understand the effect of single factor and combination of several factors in one experiment. As such, separate ad-hoc experiments can be avoided.
- Models will indicate priorities for applied research
- It will help managers in making suitable decisions.

CROP WEATHER CALENDER (CWC)

Indian Agriculture is very closely linked with the weather. The vagaries of the Indian monsoon seriously affect the crop yield. Farmer's weather Bulletins are being issued by the forecasting centers of the IMD, with an emphasis on those aspects of weather which is likely to affect the crops. To enable easy forecasting, India Agricultural Meteorology Division has prepared guides called crop weather calendars. More than 500 CWC have been prepared for important crops grown in different districts. These set out the weather situation for which warnings are to be issued during various phases of crop growth.

CWC is the pictorial form of chart containing the different stages of crops, favourable and unfavorable weather conditions prevailing during different stages of crops. CWC diagram helps any one to see at a glance what warnings are to be issued to a particular area for a given weather situation and stage of the crop.

The crop-weather calendar consists of three parts. At bottom is shown diagrammatically the different phases of the crop and the duration in days. The middle part shows the normal weather requirement. The upper most contains the warnings and the periods during which they are to be issued by the forecasting centre concerned.

Chapter - 19

Agroclimatic Normals for Field Crops

Crop yield is influenced by genetical (heredity), external (environmental) and management factors that occur during the crop growing period. The external environment is the climate which regulates and determines the growth and development and final output of crop plants. But, man has no control over weather, hence it is dominance over the success or failure of agricultural enterprises. According to the World Meteorological Organization, the weather induced variability in crop yield is as high as 50 per cent. Therefore, weather should be taken as one of the inputs in agricultural planning.

Under optimum climatic conditions, the plants manifest their maximum growth and production. Different crop growth cycles demand different climatic condition for fulfillments. Experimentally determined optimum climatic parameters for maximum production of a crop are not available in plenty as these require costly and time consuming studies employing phytotron (a green house where every growth factor can be controlled) and growth chambers. Alternatively, the climate of a region where a particular crop is best grown with less pest and disease incidences can be studied elaborately.

DEFINITION

Climatic normal means the degree of temperature, amount of rainfall, humidity etc., which distinguish optimal conditions from those defined as abnormal, both because of excess and insufficiency.

Uses of study of Agro-climatic normals for field crops can be as follows:

- Useful for Agricultural Planning.
- Useful in introduction of any crop, if the climate in which a crop is introduced, matches to the requirements of the crops, then the benefit will be the maximum.

Examples:

a) Introduction of groundnut in peninsular India from Africa.

b) Long grained rice into California. Will be useful to forecast the abnormal weather.

CLIMATIC NORMALS FOR CROP PLANTS

1. Rice

Besides rainfall, temperature and solar radiation influences rice yield, directly affecting the physiological processes involved in grain production and indirectly through the incidence of pests and diseases.

Temperature: The difference in yield is mainly due to temperature and solar radiation received during its growing seasons. It requires high temperature, ample of water supply and high atmospheric humidity during growth period. This crop tolerates upto 40^{0}C provided water is not limiting. A mean temperature of 22^{0}C is required for entire growth period. If night temperature drops lower than 15^{0}C. during flowering phase, the rice yield is greatly reduced by formation of sterile spike lets. The period during which low temperature is most critical is about 10-14 days before heading.

The optimum temperature requirements for the different stages of rice crop are given below:

Growth Stage	Temperature (^{0}C)		
	Low	High	Optimum
Germination	10	45	20-35
Seedling establishment	12-13	35	25-30
Rooting	16	35	25-28
Leaf elongation	9-18	38	31
Tillering	9-16	33	25-31
Panicle initiation	15-20	38	33
Anthesis	22	35	30-32
Ripening	12-18	30	20-25

In Tamil Nadu, kuruvai paddy yields more than samba / thaladi paddy. Because, the prevailing climate during kuruvai season favoured with high temperature and high atmospheric humidity where the irrigation water is not limited.

Solar Radiation: Low sunshine hours during the vegetative stage have slight ill effect on grain production, whereas the same situation during reproductive stage reduce the number and development of spikelets and thereby the yield. For getting higher grain yield of 5t/ha, a solar radiation of 300 cal/m^2/day is required. A combination of low daily mean temperature and high solar radiation during reproductive phase has given higher yield.

Rainfall: Rice requires high moisture and hence classified as hydrophytes. Rice requires a submerged condition from sprouting to milky stage. The moisture requirement is 125 cm. An average monthly rainfall of 200mm is required to grow low land rice and 100mm to grow upland rice successfully.

2. Wheat

Temperature: Optimum temperature for sowing is 15-20^{0}C. At maturity, it requires 25^{0}C. At harvest time, wheat requires high temperature of 30-35^{0}C and bright sunny period of 9-10 hours.

Moisture: One hectare of wheat consumes about 2500-3000 tonnes of water. Water deficiency at the heading stage results in shriveled grains and low yield. In Punjab, 35 to 40 cm of well distributed rainfall in the entire crop season or four irrigations, one at crown initiation stage and subsequently three at 40 days interval, result in good yield of wheat.

3. Maize

This crop is best suited for intermediate climate.

Temperature: Maize requires a mean temperature of 24^{0}C and a night temperature above 15^{0}C. No maize cultivation is possible in area where the mean summer temperature is below 19^{0}C, or where the average night temperature during the summer falls below 21^{0}C. However, high night temperature also result in less yield. The crop gave 40 per cent lesser yield at 29^{0}C night temperature as compared with 18^{0}C.

Moisture: Maize is adapted to humid climate and has high water requirements. It needs 75 cm of rainfall during its life period. The average consumptive use of water by maize is estimated to range between 41 and 64 cm. From germination to earing stage, maize requires less water. However, at flowering it requires more water and the requirement reduces towards maturity.

4. Groundnut

It is a tropical crop distributed between 45^0N and 30^0S latitude.

Temperature: It can be raised under a wide range of temperature. However, both very high and very low temperatures adversely affect it. A temperature range of 14-16^{0}C is necessary for the seeds to germinate. Higher temperature results in better performance in terms of length of stem, number of flowers and the number of pods. Maximum number of pods have been harvested at a mean soil temperature of 23^{0}C. The number of pods decrease as the temperature increases.

Moisture: An ideal rainfall consists of 75-125 mm during summer months preceeding sowing, 125-175 mm during a fortnight after sowing and 370-600 mm of well distributed rainfall during the crop growth.

5. Cotton

It is a hot season crop. It requires 4-5 months of uniformly high temperature (28-45^{0}C) during the crop growth period.

Mean air temperature of 21 to 29^{0}C is required at vegetative period. The optimum air temperature for reproductive phase is 27-32^{0}C; mean sunshine hours is 8-9 hrs/ day; and mean RH is 70 per cent. But at boll development and boll opening period (September to November) RH less than 70 per cent and 8 hrs of sunshine are ideal for good cotton production.

The growth rate of cotton crop is increased at 25-30^{0}C. Temperature below 15^{0}C retards the growth and reduces the square (bud) formation.

Moisture: The minimum rainfall required for cotton is 500-6500mm. Heavy rainfall during early stage is undesirable. Dry autumn months are desirable for good quality produces. Excess rainfall at later stage may cause shedding of leaves, squares and bolls. It also stimulates top growth and delays maturity and discolours lint. High humidity favours many pests and diseases.

6. Sugarcane

i. Mean air temperature for optimum germination is 30^{0}C.

ii. Mean air temperature for optimum growth is 35^{0}C.

iii. At temperature less than 20^{0}C growth is reduced.

iv. Ideal climate is 4-5 months of hot period with temperature of 30-35^{0}C followed by 6-8 weeks of cooler period for better maturity.

CHAPTER - 20

Monsoons of India

The earth has two motions. (1) Rotation and (2) Revolution. The earth rotates about its axis (west to eastward) with a speed of 464 m/s or 1669 km/hr. Its angular speed of rotation is about 7.29 x 10^{-5} radians/ sec or 0.26244 radians per hour. This rotation causes day and night and diurnal variations of weather.

The rotating earth revolves round the sun in an elliptical orbit (with the sun at one of its Foci) with a speed (orbital speed) of about 1.073 x 10^5 km/hr or 29770 m/s or 29.8 km/s. This revolution causes seasons. It has been stated earlier that the apparent path of the sun with respect to the celestial sphere is called ecliptic.

Equinoxes: When the sun apparently crosses the celestial equator (ecliptic intersects the celestial equator), all the places on the earth experience equal (length) day and nights. Every year equinoxes occur on March 21 and September 23.

Solstices: When rhe sun moves over the Tropic of Cancer (lat 231/ 2^0 N, farthest north) it is called summer solstice (June 21), which is associated with the longest day and shortest night in the northern hemisphere. When the sun moves over to the Tropic of Capricorn (lat 23 ½ ^{0} S, farthest south) it is called winter solstice (December 22), which is associated with the longest day and shortest night in southern hemisphere. The two points on the ecliptic which are farthest from the celestial equator are solstices.

Solar day: The time period (t1) between two successive crossing of the sun across the meridian (over head) varies from day to day through out the year. The average of this interval $t = \sum_{i=1}^{365} t_i / 365$ is called mean solar day = 24 hrs. Thus a sidereal day is shorter than the mean solar day by three minutes and 56 seconds.

Monsoon represents one of the phenomena in the category of secondary circulation of the atmosphere. The term monsoon is derived from an Arabic word 'Mausim' from Malayan word 'monsin' which means 'season'. The word monsoon is applied to such a circulation which reverses its direction every six months i.e. from summer to winter and vice-versa.

ECONOMIC IMPORTANCE OF MONSOON

The economic significance of monsoon is enormous, because a population of more than 2000 million lives i.e. roughly about half the world's population (54 per cent) depends on the monsoon rains for their crops. Moreover, a large percentage of total population in the monsoon region derives its income from agriculture. In India, monsoon means life-giving rains. Rice is their major crop which provides food for millions of people, hence monsoon rains are so essential for its growth. Failure of monsoon rains cause loss of food crops. Erratic behaviour of monsoon causes disastrous floods in some parts of the country, while, in other parts there is severe drought.

High and low pressure areas are developed due to differences in the amount of solar radiation received, differences in absorption of heat by water and Earth and rotation of the Earth. More solar radiation is received near the equator compared to poles. Therefore, temperatures are high near the equator are called **doldrums.**

The air in the polar region swing to the temperate zones by the rotation of the Earth. Due to the movement of air from the polar region, pressure decrease at polar region known as **polar calms**. Thus on Earth, there are two distinct low pressure areas: doldrums and polar calms. The latitudinal belt between 30 to $35^0$of both South and North are called **horse- latitudes**. The air currents at the upper layers from both the equator and poles meet at horse latitudes and produce a belt of high pressure. From these, horse latitudes winds blow towards equator and poles. Due to higher temperature near the equator, air moves upwards creating low pressure belts. The air from lower layers of atmosphere moves from both South and North latitudes to fill the low pressure belt. This type of movement of air occurring through out the year are known as **trade winds** or **tropical easterly**. They have the same course or track, year after year.

Due to higher temperature near the equator, air rises above resulting in high pressure area in the upper layers of the atmosphere which force the air to move South-wards or North-wards from the equator. These winds are called **anti-trade winds**. Direction of the anti-trade winds is quite opposite to the trade winds but occur at higher layers of the atmosphere.

South - Westerly on-shore winds blowing towards the centre of low pressure over northern India traverse thousands of miles over the warm tropical ocean. They are, therefore, full of moisture and have a great potential for heavy precipitation. The South-West monsoon, as it is called in this region, is split into two branches by the shape of peninsular India. They are known as (a) the Arabian Sea Branch, and (b) the Bay of Bengal branch.

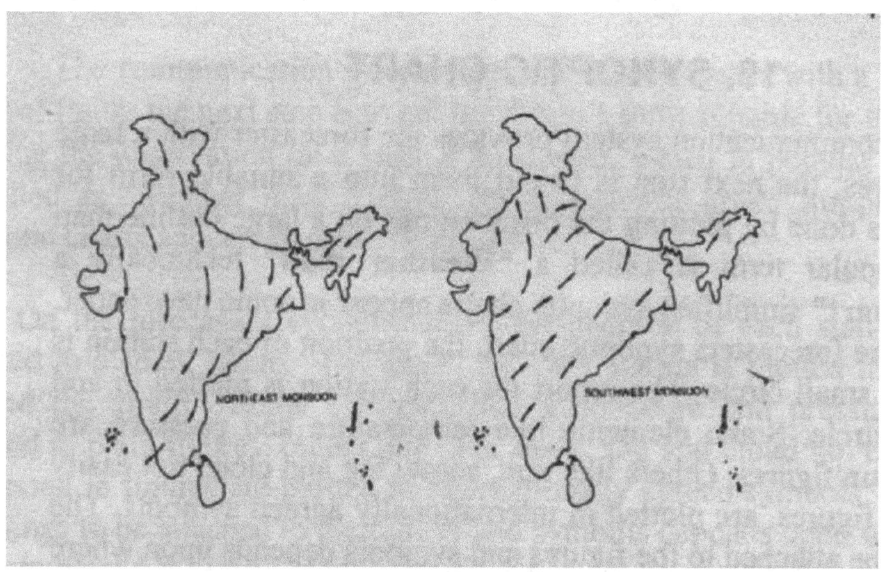

ARABIAN SEA BRANCH

The Arabian sea branch strikes the elevated Western Ghats receive heavy orographic precipitation. However, the westerly current from the Arabian Sea continues its journey across the Indian Peninsula, but the amount of rainfall on the leeward side goes on diminishing with increasing distance from the sea coast. The Western Ghats have 100-250cm of rainfall on their windward slopes, while there is well-marked rain-shadow to the leeward. Towards the North, where Western Ghats are not very high, the difference in the amount of rainfall between the windward and leeward side is rather negligible. Some of the air currents from the Arabian Sea branch manage to proceed towards Chotta Nagpur plateau through the Narmada and Tapti gaps. These air currents ultimately unite with the Bay of Bengal branch.

BAY OF BENGAL BRANCH

One current of the Bay of Bengal branch, which is more Southernly, moves towards Assam where Mausinram (near Cherapunji), situated in the southern slope of Khasi hills, has the unique distinction of recording the highest annual average precipitation (965 cm) in the world. This is because of its peculiar geographical location. A current of the Bay of Bengal branch recurves westward and advances upto the Gangetic plain. Monsoon current takes place around the eastern end of a trough winds, is of course, parallel to the Himalayan Ranges. The rainfall and partly by the cyclonic storms or monsoon depressions which followed the track of low relief and low pressure along Southern monsoon current blows from a South Easterly direction. The rainfall decreases from East to West and from North to South. The main reason why the amount of rainfall decreases Westwards is the increasing distance from the Himalaya which cause the forced ascent of rain-bearing air current.

WINTER MONSOON

A Secondary high pressure system develops over Kashmir and the Punjab. The high pressure area control the prevailing wind direction over the rest of the sub-continent. Contrary to the pressure condition over land, there are low pressure centers formed over the Indian Ocean, the Arabian Sea, and Northern part of Australia. In the cold season, therefore, there is a pressure gradient from land to sea as a result of which winds begin to move from land to sea. These are the North East or Winter monsoons of Northern hemisphere.

The Southern part of Indian Peninsula receives rainfall from North East monsoon currents. These currents while traveling over the Bay of Bengal pick up moisture from warm ocean surface. The amount of Winter rainfall on the Eastern side of the peninsula is much heavier than that on the other side. It is also known as **'retreating monsoon'**.

MONSOONS

The trade winds change their course during a part of the year, only in certain regions of the world, where local factor predominate and influence the direction of the wind. Winds thus changing their course and direction periodically are known as '**monsoons'.** India has a well-developed regular monsoon system.

INDIAN MONSOONS

India is situated in the NE trade wind zone and should have NE trade winds throughout the year, but has a regular monsoon system developed as a result of certain physiographic features. Central Asia lying North of India is a large arid surface that gets heated up during Summer when the Sun moves northward to the Tropic of Cancer and becomes a low pressure area. The Indian Ocean to the South of India is not heated upto the same extent and serves as a centre of high pressure. The SE trade wind crosses the equator, when it is deflected by the rotation of the Earth and becomes the SW monsoon wind. It gets charged with moisture, when it passes over the Indian Ocean. Clouds are consequently formed, thunderstorms develop and the monsoon bursts into rain on reaching the West coast of India by the end of May. The SW monsoon wind enters India, both from the Arabian Sea and the Bay of Bengal. The Arabian Sea branch is the most important of the two for South India. It appears in the West coast by the end of May and spreads northwards and North-eastwards. Precipitating rains over a large part of the Tamil Nadu and Andhra States. The West coast districts in Kerala get very heavy rains, with Nilgiris also getting a fair share. Regions on the leeward side of the Western Ghats like Coimbatore get very light rains and these are called **rain-shadow regions**.

The Bay of Bengal branch benefits the East coast and the Northern coast get fairly heavy rains. The Southern coastal districts in Tamil Nadu also get fair amounts of rain. Coimbatore and the southern districts, Madurai, Ramanathapuram and Tiruvelveli etc., are not benefited appreciably by this monsoon.

The SW monsoon is followed by the NE monsoon towards the end of September. The NE monsoon is also known as the retreating monsoon. This is an example of transfer of direction of the wind with the migration of the Sun southward. In the cold season from September to December, Central Asia becomes the hot centre of low pressure and draw the air from Central Asia. The wind coming from Central Asia passes over Tibet, India and the Indian Ocean to the southern hemisphere. The NE monsoon is a dry wind system. But cyclones develop at the head of the Bay of Bengal, which cross over peninsular India; Tamil Nadu, Andhra Pradesh, West Bengal and Thailand are affected frequently by such cyclones. Almost the entire Andhra Pradesh and Tamil Nadu States get a fair amount of rainfall, and the southern districts of Tamil Nadu not subject to the influence of the SW monsoon, receive good rains during the NE monsoon.

Prediction of Monsoon in India

Every year, the nation holds its collective breath as the venerable India Meteorological Department (IMD) issues its forecast of how the South-West monsoon is likely to turn out. Such concern is a reflection of how much the monsoon continues to dominate the Indian economy and rules the lives of people across this vast and diverse land. What everyone is considered to have ended in a drought when the rainfall received countrywide is less than 90 per cent of the long term average. In the last 130 - odd years, such droughts have occurred in just 21 years. However, an analysis published by Indian scientists recently showed that in the current state of the economy even a moderate drought (when the rainfall deficit ranges from 10 per cent to 15 per cent) can reduce the GDP by Rs.50,000 crore or more and slash food grain production by about 10 million tones. For the forthcoming monsoon, the IMD has predicted that the countrywide rainfall would be just five per cent below the long-term average which would classify the monsoon as a 'normal' one. Will the technique be able to give advance warning of a drought? It is believed that the new method could have successfully predicted the drought of 2004 and given at least an early indication of trouble during the 2002 monsoon, which failed badly and produced one of the worst drought in a century.

An issue of practical concern is that rainfall during the monsoon is unevenly distributed in time and across the regions. Some places can be subjected to bouts of heavy rain while other parts of the country receive hardly any rain at all; these are aberrations that average do not capture. The IMD is receiving a long overdue up gradation of its instrumentation, with plans to install a network of automatic rain gauges, wind-measuring devices, and advanced weather radars. Timely availability of data from these instruments can help to greatly increase the accuracy with which dynamical models that simulate processes in the atmosphere and ocean are able to predict the course of the monsoon a few days in advance. To further increase the lead time for forecasts, however, the dynamical models will themselves have to be improved.

But, such improvements will be possible only if there is a better understanding of factors influencing the monsoon. The monsoon's links to the oceans surrounding India are, for instance, just getting recognized. Investing in science that will increase our understanding of the monsoon holds the key to better forecasts that will immensely benefit the people of India.

THE SECONDARY WIND CIRCULATION

The general circulation of wind has already been considered. On this are superimposed local circulations of a temporary nature, caused by various external factors. These secondary circulations are responsible to an extent for the production of disturbed weather conditions like storms, cyclones, etc.

DEPRESSIONS OR CYCLONES

The isobars on local weather maps do not tend to be even roughly parallel as in the case of charts of the world's general wind circulation. Isobars assume various forms in local maps and many irregularities may often be seen. The two patterns that are of importance are the cyclonic depressions and anti-cyclones.

In the case of cyclones, a series of roughly oval isobars enclose an area of low pressure, with the pressure decreasing from the periphery to the centre; there is a gradual reduction of pressure from the rim to the centre. The size of this individual low pressure area may be 150 to 3000 km across. The depressions in the various areas may be of different patterns and they may move in different directions. The general direction of movement may be West to East, with a tendency to deflect a little in a southerly or northerly direction. These depressions have no fixed paths, though certain general tracks may ordinarily be seen. Such depressions start from the Bay of Bengal during September to November and move across peninsular India, with heavy downpour of rain and strong winds accompanied by squally weather. The Coramandel coast in the South and the Circars Coast in the North often experience cyclones during the NE monsoon season.

ANTI-CYCLONES

This has the condition of the cyclone reversed, with high pressure in the centre and low pressure at the rim. The isobars enclose larger areas than in the case of cyclones and are farther apart. While the cyclone is caused by the removal of air through the top from the trough or central basin, air is being added from the top in the case of anti-cyclones and the air flows out through the bottom.

THUNDER - STORMS

These are local phenomena, very prevalent in the tropics during the post and pre-monsoon periods. The weather is hot and sultry, thick clouds gather rapidly during the afternoon and there are squally winds and sudden downpours of rain, for a short time. Considerable lightning and thunder are noticed before and during the rains. Rains stop suddenly, the weather clears and the temperature rises again.

THE COLD WEATHER PERIOD

This period is practically rainless, though there may be very light drizzles now and then in some years. This facilitates the harvest of both dry and wet land crops of the previous season. When there are any rains, they benefit the cold weather crops like sesamum, ragi and pulses, but they interfere with the harvest of rice, sugarcane etc., and do more harm than good. Cotton plants shed their bolls, the quality of the maturing tobacco is affected adversely by the gums and resins on the surface being washed off and ripe chilli fruits cannot be dried properly. The weather during this period is usually cool, dry and pleasant, with dew in the nights. There are steady North East winds, which gradually change their direction and become easterly winds and later South easterly winds, which go by the name of *pyru galli* in Andhra Pradesh and *Uppam kathu* in Tamil Nadu. The South East wind invigorates the standing crops, particularly the cotton that is grown in dry lands.

THE HOT WEATHER PERIOD

The weather gets hotter steadily from about the beginning of March. April and May are the hottest months of the year. Some rains may be had during this period, which are called '**summer showers**' or '**mango showers**'. They are of the nature of thunder storms, often confined to the afternoons, with thunder and the lightning. Sesamum and early sorghums are sown with the rains. The standing garden land crops are benefited. The hot weather rains are confined to April and May in Tirunelveli and Ramanathapuram Districts of Tamil Nadu, which may sometimes be pre-monsoon showers. The Summer rains are beneficially made use for the preparation of the land by ploughing. The rains are uncertain in quantity and time, and cannot otherwise be utilized. The first few centimeters of soil are moistened by the rains and the prevalent high temperature evaporates this moisture quickly. It is not stored in the soil to supplement the stock of moisture in the lower layers and does not, therefore, become useful to the next crop.

THE SW MONSOON SEASON

This extends from June to September and is the grand period of general rainfall in South India. The SW monsoon rains range from 27 to 63 cm in most of the districts and from 220 to 330 cm in the West coast. Coimbatore and the southern districts of Tamil Nadu namely Madurai, Ramanathapuram and Tirunelveli alone receive a low rainfall of 8.0 to 27.0 cm. The driest part during this period is the tip at the southern end.

The SW monsoon establishes along the West coast by about the first week of June. Rains are received in other districts in July only, June being almost dry. The bulk of the sowings in the light red soils, occupying nearly two-thirds of the cultivated area, is done during this period. If the rains are delayed, sowings become unseasonal and shorter duration varieties or less valuable crops have to be sown. The delay in rains does not however affect the heavy black soils, which are usually sown late in September or October, during the NE monsoon season. The South-westerly wind gives place to North-easterly winds by about the middle of September or a little later.

The SW monsoon rains, received in the Western Ghats, feed the rivers Godavari, Krishna, Cauvery and Tamiraparani which traverse the peninsula West to East and provide irrigation for rice crops in wet lands. These rains are useful for garden land also and irrigation for standing crops is considerably reduced. The water-supply improves in under ground wells.

It may be said that the prosperity of South India is largely dependent on the reasonableness and adequacy of rains during the SW monsoon season. Most dry lands and wet lands depend directly on these rains. The garden lands are also benefited to a certain extent.

THE NE MONSOON SEASON

The period of NE monsoon extends from October to December. Heavy rains are received along the Coromandal coast. The NE monsoon wind commences about the middle of September and both SW and NE winds blow irregularly, during the second fortnight, each blowing for a part of the day. SW winds cease completely by the end of September and NE winds get established fully. The southern tracts least affected by the SW monsoon are most benefited by the NE monsoon.

The NE monsoon rains are helpful in filling the rain-fed tanks, on which rice cultivation depends and they are of special importance in tank-fed areas like Chengalpet, Ramnad, Tutukudi, Virudhunagar and Tirunelveli Districts of Tamil Nadu. Sowing in rainfed heavy black soils are done after the commencement of these rains.

The NE monsoon rains are heavy and destructive in certain years. Cyclonic storms develop in the Bay of Bengal and strike the East coast. Considerable damage is sometimes brought about by the tidal waves, flowing over cultivated lands, inundating them and making them unproductive. Heavy rains tend to lodge the rice crops during flowering stage and reduce the yield eventually. Sugarcane fields are water-logged and have to be propped up, to prevent the lodging of the canes in the North coastal districts. Tobacco nurseries and fields suffer. The harvest of early rice crops is also affected by late heavy rains. The rains benefit the southern districts effectively.

INDIA'S TREND IN RAINFALL PATTERN

Kumar and his co-workers have studied the India's rain fall for 135 years, starting from 1871 to 2005 by dividing the entire country as 30 sub-devisins.

Although the area of climate change is vast, the changing pattern of rainfall is a topic within this field that deserves urgent and systematic attention, since it affects both the availability of freshwater and food production. Based on experimentation at New Delhi, India, Aggarwal (2007) has reported that a 1°C rise in temperature throughout the growing period will reduce wheat production by 5 million tonnes. The global average precipitation is projected to increase, but both increases and decreases are expected at the regional and continental scales. Higher or lower rainfall, or changes in its spatial and seasonal distribution would influence the spatial and temporal distribution of runoff, soil moisture and groundwater reserves, and would affect the frequency of droughts and floods. Further, temporal change in precipitation distribution will affect cropping patterns and productivity.

According to the Intergovernmental Panel on Climate Change (IPCC, 2007), future climate change is likely to affect agriculture, increase the risk of hunger and water scarcity, and lead to more rapid melting of glaciers. Freshwater availability in many river basins in India is likely to decrease due to climate change. This decrease, along with population growth and rising of standard of living, could adversely affect many people in India by the 2050s. Accelerated glacier melt is likely to cause an increase in the number and severity of glacier melt-related floods, slope destabilization and a decrease in river flows as glaciers recede. Kalra *et al.* (2008) found that the yield of wheat, mustard, barley and chickpea show signs of stagnation or decrease following a rise in temperature in four northern states of India.

Under the conditions of skewed water availability and its mismatch with demand, large storage reservoirs may be needed to redistribute the natural flow of streams in accordance with the requirements of a specific region. The general practice of designing a reservoir is based on the assumption that climate is stationary. Changes in rainfall due to global warming will influence the hydrological cycle and the pattern of streamflows. This will call for a review of reservoir design and management practices in India.

The Indian climate is dominated by the southwest monsoon. About 80 per cent of the rainfall in India occurs during the four monsoon months (June–September) with large spatial and temporal variations over the country. Such a heavy concentration of rainfall results in a scarcity of water in many parts of the country during the non-monsoon period. Therefore, for India, where agriculture has a significant influence on both the economy and livelihood, the availability of adequate water for irrigation under changed climatic scenarios is very important. The agricultural output is primarily governed by timely availability of water. In future, population growth along with a higher demand for water for irrigation and industries will put more pressure on water resources.

With the growing recognition of the possibility of adverse impacts of global climate change on water resources, an assessment of future water availability at various spatial and temporal scales is needed. It is expected that the response of hydrological systems, erosion processes and sedimentation could significantly alter due to climate change. An understanding of the hydrological response of a river basin under changed climatic conditions would help solve problems associated with floods, droughts and allocation of water for agriculture, industry, hydropower generation, domestic and industrial use. Scenarios of changes in runoff and its distribution depend on the future climate scenarios.

In India, attempts have been made in the past to determine trends in the rainfall at national and regional scales. Most of the rainfall studies were confined to the analysis of annual and seasonal series for individual or groups of stations. A much wider view has been taken, and changes in rainfall have been studied on seasonal and annual scales. Intra-seasonal variability in rainfall has also been studied by analysing the trends in monthly rainfall.

Kumar *et al.* (2010) have carried out review of previous workers and taken the rain fall analysis for India for 135 years, starting from 1871 to 2005. They stated that several investigators have shown that the trend and magnitude of warming over India/the Indian sub-continent over the last century is broadly consistent with the global trend and magnitude (Hingane, 1995; Pant & Kumar, 1997, Arora *et al.*, (2005), Dash *et al.*, (2007). Pant & Kumar (1997) analysed the seasonal and annual air temperatures from 1881–1997 and have shown that there has been an increasing trend of mean annual temperature, at the rate of 0.57°C per 100 years.

Some past studies relating to changes in rainfall over India have concluded that there is no clear trend of increase or decrease in average annual rainfall over the country. Though no trend in the monsoon rainfall in India is found over a long period of time, particularly on the all-India scale, pockets of significant long-term rainfall changes have been identified.

A change detection study using monthly rainfall data for 306 stations distributed across India was attempted by Rupa Kumar *et al.* (1992). They showed that areas of the northeast peninsula, northeast India and northwest peninsula experienced a decreasing trend in summer monsoon rainfall. A widespread increasing trend in monsoon rainfall over the west coast, central peninsula and northwest India was also reported. The decreasing trend ranged between "6 and "8 per cent of the normal per 100 years, while the increasing trend was about 10–12 per cent of the normal per 100 years. Srivastava *et al.* (1998) supported the existence of a definite trend in rainfall over smaller spatial scale. Sinha Ray & De (2003) concluded that all-India rainfall and surface pressure shows no significant trend, except for some periodic behaviour. According to Sinha Ray & Srivastava (1999), the frequency of heavy rainfall events during the southwest monsoon has shown an increasing trend over certain parts of the country, whereas a decreasing trend has been observed during winter, pre-monsoon and post-monsoon seasons. These authors tried to attribute this variation to dynamic and anthropogenic causes. The inter-annual and decadal variability in summer monsoon rainfall over India was examined by Kripalani *et al.* (2003) by using observed data for a 131-year period (1971–2001). They found random fluctuations in annual rainfall and distinct alternate epochs (lasting approximately three decades) of above- and below-normal rainfall for decadal rainfall. They also concluded that this inter-annual and decadal variability appears to have no relationship to global warming. Analysis of rainfall data for the period 1871–2002 indicated a decreasing trend in monsoon rainfall and an increasing trend in the pre-monsoon and post-monsoon seasons (Dash*et al.*, 2007).

Mirza *et al.* (1998) carried out trend and persistence analysis for the Ganges, Brahmaputra and Meghna river basins. They showed that precipitation in the Ganges basin is, by and large, stable. Furthermore, one of three sub-divisions of the Brahmaputra basin shows a decreasing trend, while another shows an increasing trend. Singh *et al.* (2008) studied the changes in rainfall in nine river basins of northwest and central India and found an increasing trend in annual rainfall in the range of 2–19 per cent of the mean per 100 years.

Recent studies showed that, in general, the frequency of more intense rainfall events in many parts of Asia has increased, while the number of rainy days and total annual amount of precipitation has decreased. Goswami *et al.*(2006) used daily rainfall data to show the significant rising trends in the frequency and magnitude of extreme rain events, and a significant decreasing trend in the frequency of moderate events over central India during the monsoon seasons from 1951 to 2000. The frequency of heavy rainfall events during the monsoon season was found to be increasing over the Andaman and Nicobar islands, Lakshadweep, the west coast and some pockets in central and northwest India, whereas it was found to be decreasing in winter, pre-monsoon and post-monsoon seasons over most parts of India (Dash *et al.*, 2007). Mall *et al.* (2007) inferred that there has been a westward shift in rainfall activity over the Indo-Gangetic Plain region. An increase in intense rainfall events leads to more severe floods and landslides. The number of cyclones originating from the Bay of Bengal and the Arabian Sea has decreased since 1970, but their intensity has increased (Lal, 2001). Moreover, the damage caused by intense cyclones has risen significantly in India. In three consecutive years since 2002, there were large floods in the northeastern states of India, on 26–27 July 2005, a record 944 mm of rain fell in Mumbai, but the seasons of 2006 and 2007 saw deficient rainfall. Severe floods were observed in many parts of Gujarat and Rajasthan during the monsoon seasons of 2006 and 2007 (India Meteorological Department, 2006, 2007).

Significance of Trend

The results of the Mann-Kendall test, applied to ascertain the significance of trends in monthly rainfall indicated that, during the non-monsoon months, the increasing rainfall observed in Punjab in March; East Uttar Pradesh and Haryana in April; sub-Himalayan West Bengal, Marathwada and Telangana in October, and sub-Himalayan West Bengal in November. During the monsoon months of June, July, August and September, significant trends (both positive and negative) were detected for sub-divisional rainfall. Significant decreasing trend was detected for: sub-Himalayan West Bengal, West Madhya Pradesh and Kerala during June; for East Madhya Pradesh, Vidarbha and Chattisgarh during July; for Assam, Nagaland, Manipur, Mizoram & Tripura during August; and for Marathwada, Vidarbha, and Telangana during September. An increasing trend for sub-Himalayan West Bengal in July, Haryana, Konkan & Goa, Madhya Maharastra, Telangana, and Coastal Karnataka in August; and Gangatic West Bengal in September was found significant. Monsoon rainfall indicated a negative significant trend in East Madhya Pradesh, Chattisgarh, and Nagaland Manipur Mizoram & Tripura and an increasing significant trend only in Punjab. Assam & Meghalaya, sub-Himalayan West Bengal & Sikkim, West Uttar Pradesh, Punjab, Marathwada, Telangana and Kerala indicated a significant increasing trend in post-monsoon rainfall. Out of all area studied, the annual rainfall of Haryana, Punjab and Coastal Karnataka showed a positive trend, whilst that of Chattisgarh showed a negative trend.

Monthly analysis of regional rainfall showed that only increasing rainfall in the Central North East region in April, and decreasing rainfall in the West Central region in July and in the North East Region in August. The present study has examined trends in the monthly, seasonal and annual rainfall on the meteorological sub-division scale, the regional scale, and for the whole of India. A large data set was used, consisting of 306 stations with the length of data series of 135 years. As expected, the sub-divisional rainfall trends show a large variability - nearly half of the sub-divisions have shown an increasing trend in annual rainfall and the remainders have shown the opposite trend. The maximum increase was 2.37 mm/year and the maximum decrease was "0.76 mm/year. Region-wise, annual rainfall indicated a small increasing trend for the North West and Peninsular India and a decreasing trend in the remaining three regions. On an all-India basis, the annual rainfall showed a small decreasing trend.

CHAPTER - 21

Greenhouse Gases

Naturally Greenhouse gases are cover as blanket on the Earth and kept it about 33^0C warmer than it would be without these gases in the atmosphere. This is called as the "**Greenhouse Effect**".

Over the past century, the Earth has increased in temperature by about 0.5^0C and many scientists believe that this is because of an increase in concentration of the main greenhouse gases viz., carbon dioxide, methane, nitrous oxide and fluorocarbons. People are now calling this climate change over the past century the beginning of "Global Warming".

Fears are that if people keep producing such gases at increasing rates, the results will be negative in nature, such as more severe floods and droughts, increasing prevalence of insects and disease causing pathogens, sea levels rising and Earth's precipitation may be redistributed. These changes to the environment will most likely cause negative effects on society, such as lower health and decreasing economic development. However, some scientists argue that the global warming, we are experiencing now, is a natural phenomenon, and is part of Earth's natural cycle. Presently, nobody can prove if either theory is correct, but one thing is certain; the world has been emitting greenhouse gases at extremely high rates and has shown only small signs of reducing emissions until the last few years. After the 1997 Kyoto Protocol, the world has finally taken the first step in reducing emissions.

The "greenhouse effect" is the heating of the Earth due to the presence of greenhouse gases. It is named this way because of a similar effect produced by the glass panes of a greenhouse.

Carbon Dioxide

Carbon Dioxide (CO_2) is a colorless, odorless, non-flammable gas and is believed to be the most prominent Greenhouse gas in Earth's atmosphere. It is recycled through the atmosphere by the process of photosynthesis, which makes human life possible on the earth Photosynthesis is the process of green plants and other organisms transforming light energy into chemical energy. Light Energy is trapped and used to convert carbon dioxide, water, and other minerals into oxygen and energy rich organic compounds. Carbon dioxide is emitted into the air as humans exhale, burn fossil fuels for energy, and deforest the planet. Every year humans add over 30 billion tones of carbon dioxide in the atmosphere by these processes, and it is up thirty per cent since 1750.

Sources of CO_2

Fossil Fuels were created chiefly by the decay of plants from millions of years ago. We use coal, oil and natural gas to generate electricity, heat our homes, power our factories and run our cars. These fossil fuels contain carbon, and when they are burned, they combine with oxygen, forming carbon dioxide. The two atoms of oxygen add to the total weight.

Deforestation is another main producer of carbon dioxide. The causes of deforestation are logging for lumber, pulp wood, and fuel wood. Also contributing to deforestation are clearing new land for farming and pastures used for animals such as cows. Forests and wooded areas are natural carbon sinks. This means that as trees absorb carbon dioxide, and release oxygen, carbon is being put into trees. This process occurs naturally by photosynthesis, which occurs less and less as we cut and burn down trees. As the abundance of trees declines, less carbon dioxide can be recycled. As we burn them down, carbon is released into the air and the carbon bonds with oxygen to form carbon dioxide, adding to the greenhouse effect. Every country is going on adding carbon dioxide in the atmosphere as credit deposit in the bank. Reduced withdrawal by the way of cutting / reducing the trees cover is going on increasing the carbon dioxide balance in the atmosphere. A report has said that about 860 acres of forest land, is destroyed every 15 minutes in the tropics.

Methane

Methane is a colorless, odorless, flammable gas. It is formed when plants decay and where there is very little air. It is often called swamp gas because it is abundant around water and swamps. Bacteria that breakdown organic matter in wetlands and bacteria that are found in cows, sheep, goats, buffalo, termites, and camels produce methane naturally, and release to atmosphere. Bacteria in the gut of the animal break down food and convert some of it to methane. When these animals belch, methane is released. In one day, a cow can emit 0.5 pound of methane into the air. Since 1750, methane has doubles, and could double again by 2050. Each year we add 350-500 million tones of methane to the air by raising livestock, coal mining, drilling for oil and natural gas, rice cultivation, and garbage sitting in landfills. It stays in the atmosphere for only 10 years, but traps 20 times more heat than carbon dioxide.

Nitrous Oxide

Nitrous oxide is another colorless greenhouse gas, however, it has a sweet odor. It is primarily used as an anesthetic because it deadens pain and it is inducing the sense of laughing. For this characteristic it is called "laughing gas". This gas is released naturally from oceans and by bacteria in soils. Nitrous oxide gas raise by more than 15 per cent since 1750. Each year we add 7-13 million tones into the atmosphere by using nitrogen based fertilizers, disposing of human and animal waste in sewage treatment plants, automobile exhaust, and other sources. It is important to reduce emissions because the nitrous oxide we release today will still be trapped in the atmosphere 100 years from now.

Nitrogen based fertilizer use has doubled in the past 20 years. These fertilizers provide nutrients for crops; however, when they breakdown in the soil, nitrous oxide is released into the atmosphere. In automobiles, nitrous oxide is released at a much lower rate than carbon dioxide, because there is more carbon in gasoline than nitrogen.

Fluorocarbons

Fluorocarbons are a general term form any group of synthetic organic compounds that contain fluorine and carbon. Many of these compounds, such as chlorofluorocarbons (CFC's), can be easily converted from gas to liquid or liquid to gas. because of these properties, CFC's can be used in aerosol cans, refrigerators, and air conditioners. Studies in the 1970's showed that when CFC's are emitted into the atmosphere, they breakdown molecules in the Earth's ozone layer. Since then, the use of CFC's has significantly decreased and they are banned from production in the United States, Europe and many other countries.

The substitute for CFC's is hydrofluorocarbons (HFC's). HFC's do not harm or breakdown the ozone molecule, but they do trap heat in the atmosphere, making it greenhouse gas, aiding in global warming. HFC's are used in air conditioners and refrigerators. The way to reduce emissions of this gas is to be sure that in both devices the coolant is recycled and all leaks are properly fixed. Also, before throwing the appliances away, be sure to recover the coolant in each.

Carbon Monoxide (CO)

CO is not a significant greenhouse gas, but brings about changes in concentrations of greenhouse gases by interacting with hydroxyl radicals (OH). Concentrations of CO have increased in northern high latitudes since 1850, but did not change significantly over Antarctica during the last two millennia. The annual average concentration was about 90 ppb in 1996. The annual mean concentration is high in the Northern Hemisphere and low in the Southern Hemisphere, suggesting anthropogenic emissions in the Northern Hemisphere.

Monthly mean concentrations show a seasonal variation with large amplitudes in the Northern Hemisphere and small ones in the Southern Hemisphere. This seasonal cycle is driven by variations in OH concentration as a sink, emission by industries and biomass burning, and transport on a large scale.

Nitrogen Monoxide (NO) and Nitrogen Dioxide (NO_2)

Nitrogen oxides (NOx, i.e., NO and NO_2 are not directly influence the global warming; but, indirect means. They are not greenhouse gases, but bring about changes in concentrations of other important greenhouses gases by interacting with hydroxyl radicals (OH). In the presence of NOx, CO and hydrocarbons are oxidized to produce ozone (O_3) in the troposphere, affecting, as a greenhouse gas, the Earth's radiative balance and, by reproducing OH, the oxidization capacity of the atmosphere.

Sulflur Dioxide (SO_2)

SO_2 is not a greenhouse gas but a precursor of atmospheric sulfuric acid (H_2SO_4) aerosol. Sulfuric acid aerosol is produced by SO_2 oxidation through photochemical gas -to-particle conversion. SO_2 has been a major source of acid rain and deposition throughout industrial times. Generally, in Europe SO_2 concentrations are higher in southern regions than in northern regions. The annual mean SO_2 concentrations in the central and eastern part of Europe were lower in 1997 than in the early 1990s.

- Greenhouse gases (GHGs) control energy flows in the atmosphere by absorbing infra-red radiation. These trace gases comprise less than one per cent of the atmosphere. Their levels are determined by a balance between "sources" and "sinks". sources are processes that generate greenhouse gases; sinks are processes that destroy or remove them. Humans affect greenhouse gas levels by introducing new sources or by interfering with natural sinks.
- Warmer air can hold more moisture, and models predict that a small global warming would lead to a rise in global water vapour levels, further adding to the enhanced greenhouse effect.
- Carbon dioxide is currently responsible for over 60 per cent of the 'enhanced' greenhouse effect, which is responsible for climate change.

Green house gases in earth's atmosphere are having the properties of transparent to incoming solar radiation. The air is cooler than the earth surface during incidence of falling of solar radiation. The sun light warms the earth surface and radiating waves from earth heated the air. On an average, earth radiate back to the space the same amount of energy which it gets from the sun. The wave lengths at which the sun and the earth emit are, for energetic purposes, almost completely distinct. The surface of the earth emits upwards only, the air's green house gases radiate both up- and downwards. So some infra red comes back down to earth. In the air, the content with moisture which is the dominant green house gas in earth's atmosphere gives green house effect roughly by 60-70 per cent. Next important is carbon dioxide, followed by methane, ozone, nitrous oxide.

Clouds are another big player in radiating the solar radiation. Under cloudy sky the green house effect is stronger than under clear sky. The atmospheric green house effect is being caused by absorption and re-emission of infrared radiation. When you present yourself inside the glasshouses, it retains heat much mainly by lack of convection and advection. The clouds top in the sunshine look brilliantly white, because they reflect sunlight. The out going terrestrial infrared trapped (warming) about 30 W/m^2. The solar radiation reflected back to space (cooling) nearly 50 W/m^2. The net cloud effect (cooling) is roughly 20 W/m^2.

The reflectivity or albedo of the earth is near 0.3. It means that about 30 per cent or slightly over 100 W/m^2 of the sun's incoming radiation is reflected back to space, while roughly 240 W/m^2 or about 70 per cent is absorbed. The amount of infra red going out to space corresponds to an effective radiating temperature of about -18°C. At -18°C, about 240 Watts per sq.mt. (W/m^2) of infrared are emitted. It is just sufficient to balance the absorbed solar radiation. The clouds cause almost half of earth's albedo and perhaps 20 per cent of the natural greenhouse effect. The surface's radiative heating and the atmosphere's radiative cooling are balanced by convection and by evaporation followed by condensation. When heat the water, it takes up latent heat and evaporate. When water vapour condenses, as happens in cloud formation, latent heat is released to the atmosphere.

The temperature profile of the air column of particular location varies between day and night, from winter to summer. The lower air in the troposphere i.e 10 to 15 km gets cooler with height. A typical value cited is 6.5°C cooling/ km of altitude. This is the so-called **global mean tropospheric lapse rate**. It indicates the average rate of cooling with height. The earth surface of the troposphere must warm until they emit enough infrared to restore the balance under the enhanced lapse rate. The difference between surface emission and emission to space increase means, this is due to green house effect. If the magnitude of the global mean troposhperic lapse rate drops, then the middle and upper troposphere warm and emit more infrared to space. To regain the balance, the green house effect must decline. The mean tropospheric lapse rate is a balance between many process of energy transfer, like radiation, convection, evaporation, cloud formation and large scale air motion.

The increase in CO_2 content in the air is observed from 1800. It was 280 ppmv (parts per million by volume), where as 315 ppmv in 1958 and 358 ppmv in 1994. The other green house gas viz., methane (CH_4) increased roughly from 0.8 ppmv in 1800 to more than 1.7 ppmv in 1992. Nitrous oxide (N_2O) rose from a pre-industrial level of about 0.275 ppmv to 0.310 ppmv in 1992. The resulting enhanced greenhouse effect is often expressed in terms of radiative forcing. Green house gases render the atmosphere more opaque to outgoing infrared radiation.

Each greenhouse gas differs in its ability to absorb heat in the atmosphere. Estimates of green house gas emissions are valued in units of millions of metric tones of carbon equivalents (MMTCE), which weights each gas by its Global Warming Potential (GWP). It is the ratio of global warming, or radiative forcing, both directly and indirectly, from one unit mass of a green house gas to that of one unit mass of carbon dioxide over a period of time. Usually 100 year period is used. But any period of time can be used for estimation. The global warming potential of different green house gas is presented in the Table.

S.No.	Gas	GWP
1.	Carbon dioxide	1
2.	Methane	21
3.	Nitrous oxide	310
4.	HFC-23	11,700
5.	HFC-125	2,800
6.	HFC-134a	1,300
7.	HFC-143a	3,800
8.	HFC-152a	140
9.	HFC-227ca	2,900
10.	HFC-236fa	6,300
11.	CF4	6,500
12.	C2F6	9,200
13.	C6F10	7,000
14.	C6F14	7,400
15.	SF6	23,900

Another tiny particles present in the air is aerosols. The tiny particles with size of 0.001 to 10 micrometers are suspended in 10 to 15 km of atmosphere air. This tiny particles have influence the climate directly by scattering and absorbing the solar radiation under clear sky. Indirectly it scattered the sun light back to space. It also acting as cloud condensation nuclei, it enhance the reflectivity and life-time of clouds. From fossil fuel burning, the sulfur dioxide is coming out and yielded sulfur particle after oxidation. Adding the organic and elemental carbon to the atmosphere is through burning of fossil fuels and tropical forests. Comparatively bigger size particles of aerosol with 1 to 10 micrometers is by mineral dust by wind blowing in the region of dry soil, over cultivation and soil erosion may also adding to the atmosphere.

THE IMPACTS OF GLOBAL WARMING

Melting of Glaciers

In the last 100 years alone the global warming hastened by increase in mean temperature of the atmosphere by about 0.5 to 1.0°C. This aggravates the receding of glaciers to major extent. According to the International Commission for Snow and Ice, glaciers in the Himalayas are receding faster than in any other part of the world. If it is continued as present state, likelihood of melting and disappearance of glaciers in Himalayas will take by the year 2035. Many Himalayan glaciers including Gangotri are known to be receding for the past several years. Retreating records of Himalayan glaciers indicate that Pindari glacier is retreating at the rate of 135.2 meters per year, while for Gangotri the rate is 28 meters per year. Africa's Mount Kilimanjaro has lost 82 per cent of its ice field since it was mapped in 1912. At the current rate of melting, the snows of Kilimanjaro will be gone within 15 years. The largest glacier on Mount Kenya lost 92 per cent of its volume in the past century. The Quelccaya ice cap, the largest in the tropics, has shrunk 20 per cent since 1963. One of the Quelccaya's main glaciers, named Qori Kalis, is retreating at a rate of almost a half-meter per day. The phenomenon is not confirmed to the tropics. Glaciers in Europe, Russia, New Zealand, and the United States are also melting. With more than 1000 glaciers, Alaska is warming faster than most areas. In the past two decades the massive Columbia glacier has retreated more than 12 kilometers.

According to the study, climate models generally predict amplified warming in the polar regions, as observed in Antarctica's peninsula region over the second half of the 20th century. Many reports in recent years has envisaged that human- induced global warming will cause the thinning ice-sheet to break off from the Antarctic continent and inundate the world's shorelines with a drastic rise in sea levels.

Sea level rise will result in loss of land, increased vulnerability to flooding, including storm events, accelerated erosion along the coasts and in river mouths, increased salinisation and changes in the physical characteristics of tidal rivers.

India has the coastline of about 7000 kms. These areas are comparatively fertile land and has thickly populated. Normally, this area occupied in the low-lying area. Hence, when sea level rise will occur, these areas are prone to inundated. The effect of one metre rise in sea level will affect the coastal areas of India.

State/Union Territery	Coastal area (million hectares)		
	Total	Inundated	Per centage
India	139574	0.571	0.41
Andhra Pradesh	27504	0.055	0.19
Goa	0.37	0.016	4.34
Gujarat	19602	0.181	0.92
Karnataka	19179	0.029	0.15
Kerala	3886	0.012	0.30
Maharastra	30771	0.041	0.13
Odissa	15571	0.048	0.31
Tamil Nadu	13006	0.067	0.52
West Bengal	8875	0.122	1.38
Andaman & Nicobar Islands	0.825	0.006	0.72

Salinization of the fertile cultivable area near sea will reduce the agricultural productivity and total production. There will be more evaporation from large bodies of water. This leads to more clouds and precipitation will more in certain area and in some area there will be scanty. The uncertain and unpredictable rainfall will be more frequent years leads to planning of crops and cultivation season is difficult. There may be more hurricanes in some area due to global warming. Certain diseases like vector-borne diseases and heat-stroke will become more widespread.

Researchers have estimated that only 2^0C increase in mean air temperatures will be enough to decrease rice yield by 0.75 t/ha in high yield areas like Punjab, Haryana and Utter Pardesh. Similarly, a 0.5^0C increase in winter temperature is likely to translate into a decrease of 10 per cent of wheat production in these areas. It is also estimated a drastic increase in green house gases like carbon dioxide may cause wheat production to fall as much as 68 per cent. One meter rise in sea level has the potential to cause losses of hundreds of crores of rupees. It is forecasted that nearly 7 million people may be displaced, 6000 square kilometers of land may get submerged and 4000 kilometres of road net work may be inundated. The changing climate conditions have the potentials to significantly increase tropical disturbances like cyclones and storms in coastal regions. In fact, it is estimated that the average number of tropical disturbance may jump two-folds. The enduring power shortages faced by most Indian States are likely to aggravate further as more and more people switch to using power - hungry air conditioners to beat the heat.

CONSEQUENCES CHANGE IN CLIMATIC TEMPERATURE

Unseasonal rains, debilitating drought, excessive floods, devastating cyclones and storms, all these are warning signals that a distressed Gaia is sending out to humankind. Climatologists and scientists, for their part, have been studying symptoms of climate change such as receding arctic ice caps and disappearing wildlife habitats that are not readily apparent to the rest of us. They are coming up with convincing proof that our climate is indeed changing in ways that differ from its usual cyclical behavior. And now comes the fourth assessment report of the inter-governmental panel on climate change, which conclusively links high concentrations of anthropogenic emissions to human activity.

When a somewhat similar threat-also caused by human activity-surfaced nearly three decades ago, the global community reacted with alacrity to cobble together a cohesive and co-ordinate response. Three scientists working independently linked the 'hole in the ozone layer' to CFCs (chlorofluorocarbons) from refrigeration; the scientists could show that the relationship between the hole in the ozone layer and the resultant ultraviolet radiation could lead to exponential increase in skin cancer. Alarm bells rang around the world, loudly enough to persuade countries to think and act collectively. As many as 150 countries came together to sign and ratify the Montreal protocol, which effectively caps and arrests CFC release into the atmosphere. So effective was this effort that already there are signs that the ozone hole is expanding. The ozone hole over the Antarctic had already shrunk by 20 per cent by 2004. Scientists are hopeful that the ozone layer will return to its original form in 50 years, thanks to timely intervention by humanity.

Some of the devastating consequence detailed in the provisional February 16, 2007, IPCC report on Asia: sea level will rise by at least 40 cm by 2100, inundating areas on the coastline, including some of the most densely populated cities whose populations will be forced to migrate inland or build dykes-both requiring a financial and logistical changes that will be unprecedented. In the South Asian region as a whole, millions of people will find their lands and homes inundated. Up to 88 per cent of all of Asia's coral reefs, termed the" rainforests of the ocean" because of the critical habitat they provide to sea creatures. May be lost as a result of warming ocean temperatures.

The Ganga, Brahamaputra, and Indus will become seasonal rivers, dry between monsoon rains as Himalayan glaciers will continue their retreat, vanishing entirely by 2035, if not sooner. Water tables will continue to fall and the gross per capita availability in India will decline by over one-third by 2050 as rivers dry up, water scarcity will in turn affect the health of vast populations, with a rise in water - borne diseases such as cholera. Other diseases such as dengue fever and malaria are also expected to raise greater crop loss- of over 25 per cent-as temperatures rise to up to 5.4 degrees C by the end of the century. This means on even lower caloric in-take for India's vast rural population already pushed to the limit, with the possibility of starvation in many rural areas dependent on rainfall for their crops. Even those areas that rely on irrigation will find a growing crisis in adequate water availability.

This grim future awaits India in the coming century. The irony is that much of this damage will be self-inflicted unless the country is prepared to make a radical, enlightened change in its energy and transportation strategies.

We are truly at a crossroads: Either we can be complacent or wait for leadership from a reluctant United States, the largest greenhouse gas emitter in the world, or begin to take action now, regardless of what other countries do.

The path that India has taken thus far, of waiting until wealthy countries take action on global warming, is understandable if viewed in isolation. The U.S., the U.K., and other countries in the wealthy North, have developed their economies largely thanks to fossil fuels. It is only fair that India be allowed to attain the same standard of living before curbing its emissions.

But as the IPCC report makes clear, while it may be "fair" to do so, it is also suicidal for India to pursue any strategy but the least carbon-intensive path toward its own development. Wealthy, less populous countries in the North are very likely - and very unfairly - going to suffer fewer devastating blows to their economies, and may actually benefit with extended growing seasons, while India and other South Asian nations will dramatically and painfully suffer if action is not taken now.

Today, much of India's energy comes from coal, most of it mined in the rural areas of Odisha, Jharkhand, and Bihar with devastating consequences. Tribals and small and marginal peasants are being forced to resettle as these mines grow wider by the day. Inadequate resettlement plans mean more migration of landless populations to urban slums. The environment is being destroyed by these mines and their waste products among them fly ash laced with heavy metals and other topic materials. But the biggest irony of this boom in coal-fired power is that much of the power is that much of the power is going to export-oriented, energy - intensive industry. Look at Odisha's coal belt and find a plethora of foreign-owned and India aluminums smelters, steel mills, and sponge iron factories - all burning India's coal, at a heavy cost to local populations - then exporting a good share of the final product to the China, the U.S. or other foreign markets.

Volatile Mix

Add to the problem of export - oriented. Energy - intensive industry the problem of carbon trades, and you have a volatile mix. India is one the top destinations globally in the growing carbon market. In exchange for carbon trade projects in India, wealthy polluters in the North are able to avoid restrictions on their own emissions. Rather than financing "clean development" project as promised, many of these trades are cheap, dirty, and harmful to the rural poor. Fast - growing eucalyptus plantations are displacing farmers from their land and tribal's from their forests. Sponge - iron factories are garnering more money from carbon trades earned by capturing "waste heat" then from the production of the raw material itself. Toxic fly ash from coal - fired power plants is being turned into bricks, and the carbon that would have been released from traditional clay - fired brick kilns, is now an invisible commodity that can be sold as carbon credits. These carbon trades are not helping finance clean energy and development for India's rural poor.

Add to this the special economic zones or SEZs - forcing people off their land, where blood, often of the most vulnerable, is shed at the altar of development.

Global warming will tighten this growing squeeze to a noose, as huge areas of Bangladesh go underwater and environmental refugees flood across India's borders. The leaked final draft of the IPCC report shows that Bangladesh is slated to lose the largest amount of land globally approximately 1000 square km of cultivated land - due to sea level rise. Where will all of those hungry, thirsty, landless millions go? Most will flock to the border looking for avenues to enter, exacerbating an already tense situation not only in the States contiguous to Bangladesh but in cities as far off as Mumbai and

Global warming is exacting the highest price on those least responsible for the problem. But India can show the world that there is another way forward: A self-interested, self - preserving way focused on clan energy such as solar and wind; on energy efficiency; on providing for its own population's energy needs ahead of foreign corporations; on public transportation plans that strengthen India's vast network of rail and bus transportation routes, rather than weakening it with public subsidies to massive highways and to automakers. The IPCC final draft report urges India and other Asian countries to prepare for the coming climate apocalypse with crop varieties that can withstand higher temperatures, salinated aquifers, and an increase is pests. It also advises better water resource management and better disease monitoring and control. While important, prevention is always the best medicine.

The IPCC final draft report should be seen as a conservative assessment of what lies in store. It clearly implies that incremental or palliative responses to reduce vulnerability are not the answer. India and the other countries of the region need to take a preventative approach by moving their economies away from fossil fuels and toward clean, renewable forms of energy. This is the only way of preserving a sustainable way of way of life that could be a model for the world.

ACID RAIN

Rainfall Chemistry

Acid Rain is a strange and variable mixture of a variety species. The pollutant gases - mainly oxides of sulphur, nitrogen and carbon - interact with other components of the atmosphere forming various products which are modified by time over distance. Commonly, sulphuric (70 per cent) and nitric acids (30 per cent) dominate. Ammonia from live - stock sources and agricultural fields, hydrogen chloride from sea - spray, volatile organic compourds (alkanes, polychlorinated aldehydes, esters and hydrocarbons) also add up to the 'acid rain cocktail." Hence, it is a secondary and cumulative pollutant too. The ubiquitous photo - oxidant ozone plays a catalytic role in these atmospheric transformations.

A simple model of acid rain generation envisages six stages of reactions:

- Emission of acid gases (SO_2 and NO_2) from natural and manmade sources at the ground level.
- Some oxides fall back directly to the ground around the point of emission without delay as gases and, aerosols (dry deposition).
- Formation of ozone, the photo - oxidant by sunlight.

- Interaction of ozone with SO_2 and NO_2 to produce H_2SO_4 and HNO_3 by oxidation.
- Dissolution of sulphur and nitrogen oxides, photo oxidants and other gases, including ammonia, in cloud (rain - out) or rain water (wash - out) to produce acids. In coastal regions, sulphates and chlorides from sea spray also add up.
- Settling down of acid rain containing the ionic species as fog, snow and rain at a distance far from the emission sources (wet deposition)

In 1852, the first Alkali Inspector of England Robert Angus Smith observed unusual acidity in the winter rains of Manchester city. He opened that a link between the sooty sky that transformed into a fog and the acidic nature of the precipitation. He did it correctly and the concept of acid rain as an air pollution problem was born in 1872, twenty years later, he coined the term 'acid rain' to describe that curious atmospheric phenomenon and published the first ever treatise.

The seriousness of the problem was not realized till Svente Oden, a Swedish soil scientist's sensation entry in 1960's. He was attracted by the severe acidification of Sweden's water bodies and the progressive decimation of fishes and other aquatic organisms. He found the waters extremely acidic and traced the cause to the emissions from the U.K. and Eastern Europe to Scandinavia. About 25 per cent of Sweden's water surface was either acidified or threatened at that time. He further cautioned that a central area in Europe was already suffering from acidic precipitation and expanding yearly. Outraged by the damages, Oden termed it as 'man's chemical warfare on nature' and visualized the broad ecological implications of long distance transport of air pollutants. The acid gases fumes are emitted in one country transcend the oceans and damaged the neighboring places. Realizations of this transfrontier transport nature of acid rain formation added a new twist to the issue involving neighboring nations.

Development of Organization

Svente Oden's sustained campaign on acid rain. Its assumed causes and possible effects kindled interest in an otherwise stagnant field. It was raised by Sweden at the first U.N. conference on human environment in Stockholm in 1972. Oden's report warned of a chemical desert engulfing entire Europe and other parts of the world. On the strength of his arguments, the organization of economic Co- operation and development (OECD) in Paris set up along Range Transport of air Pollutants (LRTAP) programme followed by a European Monitoring and Evaluation programme (EMEP) under the auspices of Economic commission for Europe - all to monitor sulphur deposition rates. Clearly, all acid rain problems became an international issue. Earlier, in 1971, the Swedish Ministry of foreign Affairs first accused its European neighbors of unwittingly dumping noxious oxides downward in Scandinavia.

Slowly, tension developed between the U.K. and its European neighbors; it was so between the U.S. and Canada also. Britain, the self - proclaimed leader of the world in pollution control was defiant and unrelenting much to the chagrin of its European neighbors. A majority of scientists and many Canadian and European politicians were arraigned against a caucus of political administrators of the U.K. and U.S. and a quarrelsome interest intensified.

Contributors of Pollution

In 1983, U.K. had an average monthly 'sulphur' deposition of 84,000 million tones, of which only 20 per cent came from other countries but, 92 per cent of Norway, 82 per cent of Sweden, 77 per cent of Netherlands, 64 per cent of Denmark, 58 per cent each of Belgium and Poland's sulphur deposits originated from the U.K. the per capita energy consumption of the U.S. is higher and its sulphur emissions totaled $22x10^6$ tonnes in 1985 and were projected to $27x10^6$ by the year 2000. These pollutants affected about two - thirds of land area in the U.S. itself. But it was Canada, the neighboring country which received most of the acidic gases which turned its lakes barren and forest dead.

With evidences mounting against the two superpowers as 'polluters', denials of complicity and swearing by clean - air principle became routine but without conviction. By June 1976, Britain had to confess its complicity in fouling atmosphere over Norway and Scandinavia. The wheel now has turned a full circle. Dissensions become deep rooted. And acid rain became an important item in the agenda of summit meetings and international conferences.

The trans - frontier effects of acid rain could, in part, be traced to the clean - air regulation enforced by the British pollution control Authorities. Aware of the potential of soot from chimneys forming smogs in and around industrial centers like Manchester, London and oxford, the power plant management was instructed to erect tall chimneys as per provisions of Clean Air Act. The rationale was dilution as solution to pollution. The traumatic experiences of London Smog in 1952, where 'the pestilent congregation of vapors' transformed into a suffocating acidic smog, blanketed the city for five consecutive days and caused more bronchial sufferings and death, and caused more bronchial suffering and death, had driven them to such desperate measures.

Causes of Air Pollution

Transportation is the prime contributor of NOx and the NO_2 level has doubled in the period between 1959 – 73. About 550 million motor vehicles roam around everybody accounting 40 around the world every day. From the 1980 level of 78 Tg sulphur and 21.3 Tg of nitrogen, it is likely to multiply thrice by the year 2020 and the increases are likely to be concentrated in Asia.

Injected into higher layers of the atmosphere, the pollutant gases were transported to surrounding countries and depending upon the direction. Time and speed of wind caused incalculable damages elsewhere - earning the sobriquet - 'the invisible invader', it was the result of the 'tall stacks paradox' a description of the Brstille (1981) opined 'well intentioned regulations may have unwittingly aggravated the problem'.

Till recently 'acid rain' was perceived as a phenomenon of wet deposition of dilute acids, mostly sulphuric acid (70 per cent) some nitric acid (30 per cent) and very little hydrochloric acids. But wide ranging chemical transformations take place while the acid gases pass through the atmosphere. The oxides of sulphur, nitrogen and carbon collectively called acidic gases - are emitted from burning of coal and petroleum products in industries and automobiles. They may settle down in the gaseous from and as aerosols near the emission points. This is called dry deposition. Alternatively, they may be carried further by the wind - directed plume, mixed, dispersed and photo - oxidized to the respective acidic forms. The emission plume is dispersed by the natural turbulence in the mixing layer 1-2 km above the ground. A buoyant uplift keeps it aloft and the wind carries it to long distances (5 – 25 km). If turbulence is violent, the plume assumes a conical shape, the constituent oxides mix with atmospheric air and the cone will become invisible. The flight period during which oxidation and dissolution occur may last for several days, cover hundreds of kilometers and cross national boundaries and geological barriers. The flying pollutants while settling down in far flung territories has become 'in visible invaders', when they interact with rain - formation events they may settle down as rain, fog, snow, hail and smog meaning wet depositions.

Pure rain, as it descends from the clouds is not neutral in pH, may be surprising; natural events like volcanic eruptions, lightning and forest fires release CO_2 which mixes with the rain water forming carbonic acid. This being a weak acid reduces the pH, due to pollutants from an - tropic sources of transportation and industry would be termed acidic precipitation. Through the pH of the water, as evident from analysis of Arctic ice deposits had been declining by 0.007 unit per year, it is the enormous imputs of industrial emissions, is the cause for concern.

CHAPTER - 22

Synoptic Chart

The communication system provides the forecaster with a large mass of figures; the next step is to put them into a suitable form for study. This is done by plotting the observations on a large outline map which in popular term is called a **"Weather map"** technically a **"Synoptic chart"** simplified synoptic charts appear in some newspaper.

A chart or map on which meteorological data presented and analysed is called a **synoptic chart.** Synoptic means having the same view point. The representation of the meteorological data in synoptic chart are atmospheric wind, temperature, pressure, cloud, precipitation, dew point etc., over a long area at a given local standard times.

On the forecasters synoptic chart, the position of each station is marked by a small circle. The report for each station is plotted in and around the circle. Some elements like temperature and pressure are entered in plain figures. Others like rain, snow, fog and cloud not easily expressed in figures, are plotted in internationally agreed symbols. The meanings to be attached to the figures and symbols depend upon where they are placed in relation to the station circle.

Thus the amount of shading in the circle is an indicator of the proportion of sky covered by cloud, the temperature (in whole degrees) is written to upper left of the circle, the sea level pressure in millibars and tenths to the upper right the hundreds figure for the pressure. The wind is represented by an arrow flying with the wind and drawn towards the station circle. The speed by feathers on the wind arrow, a short feather indicating 5 knots, a large one 10 knots, a long and short

15 knots and so on. The following parameters are being included in the synoptic chart.

Air temp ^{0}C.	Type of high cloud
Wind speed	Type of medium cloud
Dew point Temp ^{0}C.	Type of low cloud.
Rainfall	Pressure

When the plotting of synoptic chart is completed, the forecaster then proceeds to the analysis, the object of which is to systematize the collection of individual station plots into coherent picture. The first stage is to draw the isobars-lines along which the pressure is the same. The completed isobars usually revealed a few standard patterns, like low pressure (cyclones) high pressure (anticyclones) etc. and how they will change in due course. In India such synoptic charts area drawn two times daily (08.30 IST and 17-30 hours IST).

Symbols used in weathers maps

General weather condition	Cloud type	Cloud coverage (oktas)
Rain	cirrus	0
Continuous slight rain	Cirrocumulus	1
Continuous heavy rain	Cirrostratus	2
drizzle	altocumulus	3
snow	altostratus	4
Thunder storm	stratocumulus	5
shower	nimbostratus	6
hail	cumulus	7
fog	cumulonimbus	8
Mist Squall haze		

Wind velocity 1 knot = 1.836 km / hr.

Upto 5 knots	sky clear
Upto 10 knots	partly cloud
Upto 50 knots	overcast
West wind	rains
East wind	fog

SURFACE CHART

A synoptic chart on which surface wind (direction and speed), visibility, pressure, temperature, dew point, rainfall, cloud (type and amount, present weather (fog, mist, rain, thunderstorms, squalls, etc.,) past weather (with respect to the time of observation) are plotted is called **surface chart** or **sea level chart**. Observation of these elements recorded or observed from ground. Except the station pressure which is reduced to mean sea level (MSL) all other elements do not belong to a constant level.

❑❑❑

References

Bisnoi,O.P. 2007. Principles of Agricultural Meteorology. Oxford Book Company, Jaipur.

Chowdhary, A.,Das, H.P and Mukhopadhyay, R.K.1990. Distribution of dew and its importance in moisture balance for rabi crops in India. Mausam 41 (4):pp.547-554.

Das. P.C. 2000. Crops and their Production Technology under different conditions, Kalyani Publishers, Chennai.

Duvdevani S.1957. Dew research for arid agriculture discovery, 18, pp.330-334.

Harpal Singh Mavi, 1974. Agricultural Meteorology, Punjab Agricultural University, Ludhiana.

Horfmann,G. 1958. Dew measurements by thermo dynamical means. International Union of Geodesy and Geophysics. International Association of Scientific Hydrology, General Assembly, Toronto, Geotburge, 2.pp.443-445.

Howward.J. Critcfield 1990. General Climiatology, Fourth edition. Prentice Hall of India Private Limited, New Delhi-1.

John.G. Lockwood,1974. World climatology, An Environmental approach. The English Language Book Society and Edward Arnold Publishers Ltd.

Kumar, V., Jain, SK and Singh, Y. 2010. Analysis of longterm rainfall trend in India. Hydrol. Sci. J.55(4), 484-96.

Monteith, J.L 1963. Dew: Facts and Fallacies, The water relations of plants. Edited by A.R.Rutter and F.H.Whitelead, Blackwell Scientific Pub. London.

Muller,W.1968. The role of measurement of dew, Agroclimatological methods, UNESCO, Paris. pp.303-307.

Navale Pandharinath.2007. The Science of Weather and Environment. BS Publications, Hyderabad.

Raman, C.R.V., Venkataraman,S. And Krishnamurthy, V.1973. Dew over India and its contribution to winter-crop water balance. Agri.Met,11pp.7-35.

Ramdas, L.A. 1943. The climate of the air layers near the ground at Poona, India.Met.Dep.Tech.Note3,p.14.

Reddy. S.R. 1999. Principles of Agronomy, Agrometeorology, Kalyani Publishers, Chennai.

Sascheti, A.K.1985. Agricultural Meteorology - Instructional-cum-Practical manual. National Council of Educational Research and Training.

Tuller, S.E. and Chilton, R. 1973. The role of dew in the seasonal moisture balance of a summer dry climate. Agri Met.11pp.135-142.

Austin Bourke, P.M. 1968. Introduction: The aim of agrometeorology. In Agroclimatological methods' proceedings of UNESCO Symp. held at Reading, U.K.p. 11-16.

Sreenivasan, P.S. 1969. The Role of Agrometeorology in Scientific Farming. Agric. For. Meteoro., 38, p.259 -62

Zeitfracht Medien GmbH
Ferdinand-Jühlke-Straße 7
99095 Erfurt, Deutschland
produktsicherheit@kolibri360.de